AF843087

Modeling DBP Formation Kinetics: Mechanistic and Spectroscopic Approaches

The mission of the Awwa Research Foundation (AwwaRF) is to advance the science of water to improve the quality of life. Funded primarily through annual subscription payments from over 1,000 utilities, consulting firms, and manufacturers in North America and abroad, AwwaRF sponsors research on all aspects of drinking water, including supply and resources, treatment, monitoring and analysis, distribution, management, and health effects.

From its headquarters in Denver, Colorado, the AwwaRF staff directs and supports the efforts of over 700 volunteers, who are the heart of the research program. These volunteers, serving on various boards and committees, use their expertise to select and monitor research studies to benefit the entire drinking water community.

Research findings are disseminated through a number of technology transfer activities, including research reports, conferences, videotape summaries, and periodicals.

Modeling DBP Formation Kinetics: Mechanistic and Spectroscopic Approaches

Prepared by:
Gregory V. Korshin, Mark M. Benjamin, and **Hyun-Shik Chang**
University of Washington
Department of Civil and Environmental Engineering
Box 352700, Seattle, WA 98195-2700

Jointly sponsored by:
Awwa Research Foundation
6666 West Quincy Avenue, Denver, CO 80235-3098

and

U.S. Environmental Protection Agency
Washington, D.C.

Published by:

Awwa Research Foundation
Advancing the Science of Water®

IWA Publishing

DISCLAIMER

This study was funded by the Awwa Research Foundation (AwwaRF). AwwaRF assumes no responsibility for the content of the research study reported in this publication or for the opinions or statements of fact expressed in the report. The mention of trade names for commercial products does not represent or imply the approval or endorsement of AwwaRF. This report is presented solely for informational purposes.

Published by IWA Publishing, Alliance House, 12 Caxton Street, London SW1H 0QS, UK

Telephone: +44 (0) 20 7654 5500; Fax: +44 (0) 20 7654 5555; Email: publications@iwap.co.uk;
Web: **www.iwapublishing.com**

AwwaRF report number 91000F. Originally published by AwwaRF for its subscribers in 2003.
IWA Publishing version published 2004.

ISBN 1 84339 8958

CONTENTS

TABLES

FIGURES

FOREWORD

The Awwa Research Foundation is a nonprofit corporation that is dedicated to the implementation of a research effort to help utilities respond to regulatory requirements and traditional high-priority concerns of the industry. The research agenda is developed through a process of consultation with subscribers and drinking water professionals. Under the umbrella of a Strategic Research Plan, the Research Advisory Council prioritizes the suggested projects based upon current and future needs, applicability, and past work; the recommendations are forwarded to the Board of Trustees for final selection. The foundation also sponsors research projects through an unsolicited proposal process; the Collaborative Research, Research Applications, and Tailored Collaboration programs; and various joint research efforts with organizations such as the U.S. Environmental Protection Agency, the U.S. Bureau of Reclamation, and the Association of California Water Agencies.

This publication is a result of one of these sponsored studies, and it is hoped that its findings will be applied in communities throughout the world. The following report serves not only as a means of communicating the results of the water industry's centralized research program but also as a tool to enlist the further support of the nonmember utilities and individuals.

Projects are managed closely from their inception to the final report by the foundation's staff and large cadre of volunteers who willingly contribute their time and expertise. The foundation serves a planning and management function and awards contracts to other institutions such as water utilities, universities, and engineering firms. The funding for this research effort comes primarily from the Subscription Program, through which water utilities subscribe to the research program and make an annual payment proportionate to the volume of water they deliver and consultants and manufacturers subscribe based on their annual billings. The program offers a cost-effective and fair method for funding research in the public interest.

A broad spectrum of water supply issues is addressed by the foundation's research agenda: resources, treatment and operations, distribution and storage, water quality and analysis, toxicology, economics, and management. The ultimate purpose of the coordinated effort is to assist water suppliers to provide the highest possible quality of water economically and reliably. The true benefits are realized when the results are implemented at the utility level. The foundation's trustees are pleased to offer this publication as a contribution toward that end.

Edmund G. Archuleta, P.E.
Chair, Board of Trustees
Awwa Research Foundation

James F. Manwaring, P.E.
Executive Director
Awwa Research Foundation

ACKNOWLEDGMENTS

The authors of this report are deeply indebted to Steve Reiber, Yujung Chang, and Christine Compton of HDR Engineering, Inc., for their technical and administrative assistance throughout the project, and to HDR Engineering, Inc., for financial support. The authors are also indebted to the following water utilities and individuals for their cooperation and participation in this project:

- Metropolitan Water District of Salt Lake and Sandy, Salt Lake City, Utah, John Robert Carman
- City of Corpus Christi Water Department, Corpus Christi, Texas, Ed Garana
- Jordan Valley Water Conservancy District, West Jordan, Utah, Richard Bay
- The City of Fort Worth, Fort Worth, Texas, Dale Fisseler
- Tarrant Regional Water District, Fort Worth, Texas, David Marshall
- Metropolitan Utilities District, Omaha, Neb., Joel Christensen
- Council Bluffs Water Works, Council Bluffs, Iowa, John Elliott
- St. Paul Water Utility, St. Paul, Minn., David Schuler
- Minneapolis Water Works, Minneapolis, Minn., Larry Cole
- Sioux Falls Public Works, Sioux Falls, S.D., Randy Janssen
- City of Tampa, Tampa, Fla., Malachi Bennett III

The advice and assistance of the Project Advisory Committee—including Philip Singer, Professor of Environmental Sciences and Engineering, University of North Carolina, Chapel Hill, N.C.; Hiba Shukairy, USEPA, Cincinnati, Ohio; and Cordelia Hwang, Metropolitan Water District of Southern California, Los Angeles, Calif.—and the help of Awwa Research Foundation project officers Kenan Ozekin and My-Linh Nguyen, are appreciated.

EXECUTIVE SUMMARY

INTRODUCTION

Chlorination of drinking water has been used to control waterborne infectious disease for more than a century and is widely recognized as one of the most effective public health measures ever undertaken. Although the formation of halogen-containing disinfection by-products (DBPs) via reactions of chlorine with natural organic matter (NOM) and with inorganic bromide represents a significant public health risk and therefore places constraints on chlorination practice, the practice of water chlorination is expected to continue and possibly to expand in the future.

This report describes research directed at developing reliable and rational rate expressions for the formation of chlorinated DBPs in potable water. The majority of existing models for DBP formation are statistical rather than mechanistic, a feature that almost guarantees that they will be highly site-specific and not reliable for predictions outside of the range of existing observations. To improve upon those models, this research focused on development, calibration, and testing of a model for DBP formation based on the stepwise sequence of reactions that constitutes the current qualitative understanding of the process. The goal was to include in the model explicit equations for (1) the rate of disappearance of precursor sites and of active chlorine and bromine, (2) the rates of formation and disappearance of halogenated reaction intermediates, and (3) the rate of formation of trihalomethanes (THMs) and haloacetic acids (HAAs). The model framework was intended to be sufficiently generic to be extended to other DBPs and total organic halide (TOX) species at a future time.

An additional goal of the project was to extend the authors' prior work relating DBP formation to the change in the ultraviolet (UV) absorbance of the water when it is chlorinated, a quantity referred to as the differential absorbance, ΔA_{λ}:

$$\Delta A_{\gamma} \equiv A_{\lambda,\,t} - A_{\lambda,\,0}$$

where $A_{\lambda,t}$ and $A_{\lambda,0}$ are the absorbance of the sample at wavelength λ at some time t after chlorination and at time zero (prior to chlorination), respectively.

Because the absorbance at 272 nm always decreases upon chlorination, ΔA_{272} is always negative. With increasing chlorination dose or contact time, the decline in A_{272} increases, i.e., ΔA_{272} becomes progressively more negative. To avoid the awkwardness of describing ΔA_{272} as "increasing in a negative direction," throughout the report, trends are presented as a function of the absolute value of A_{272}, i.e., as a function of $|\Delta A_{272}|$.

METHODS AND MATERIALS

Numerous experiments were carried out with a baseline water that was collected from Lake Washington (LW), in Seattle, Wash. In these experiments, the pH, temperature, and bromide concentration of the water were altered, after which it was dosed with free chlorine. The decline of the free chlorine concentration, the change in the absorbance spectrum of the sample, and the formation of various DBPs were then analyzed after reaction times up to seven days. A unique feature of this research was the collection of data at very short reaction times (from < 1 s to 100 s) using sophisticated stopped-flow and quench-flow analytical instrumentation. The kinetics of DBP formation in several other raw waters from around the United States were also investigated, although the range of experimental conditions was less extensive in those experiments.

RESULTS

Medium- and Long-Term Kinetics of Chlorination of Lake Washington Water

In the baseline experiments with LW water, chlorine was dosed such that the ratio of the initial chlorine concentration (as Cl_2) to the DOC concentration (the Cl_2:DOC ratio) was 1.5 mg/mg. In these systems, the changes in free chlorine concentration and UV absorbance and the concentration of DBPs all followed patterns that have been observed in previous studies of the chlorination of natural waters. Specifically, these parameters underwent an almost instantaneous change (a decline in chlorine concentration and UV absorbance, and an increase in DBP concentration) when the water was dosed with chlorine, followed by a more gradual change that continued for at least several days at a steadily declining rate.

The relationship between chlorine consumption and the decline in UV absorbance at 272 nm ($|\Delta A_{272}|$) was curvilinear in a way that implied that the initial reactions of NOM with chlorine destroy more absorbance than subsequent reactions do. A similar relationship was observed between chloroform formation and $|\Delta A_{272}|$. This relationship is consistent with those observed in prior research and also with the idea that the initial reactions between NOM and chlorine destroy some UV absorbance, but that chloroform is generated only as an end product of a more extensive sequence of reactions.

All the patterns described above were followed when the chlorine dose was adjusted to achieve Cl_2:DOC ratios over the range from 0.1 to 3.0 mg/mg. The rates and ultimate extents of all three processes (Cl_2 consumption, ΔA_{272} decline, and DBP formation) increased slightly with increases in the Cl_2:DOC ratio, so that plots of Cl_2 consumption vs. $|\Delta A_{272}|$ and $CHCl_3$ concentration vs. $|\Delta A_{272}|$ were very similar for the full range of Cl_2:DOC ratios investigated. The concentrations of most other DBPs detected in these experiments ($CHCl_2Br$, dichloroacetic acid [DCAA], trichloroacetic acid [TCAA], and chloral hydrate [CH]) could also be plotted as coherent, reliable functions of $|\Delta A_{272}|$ for the whole range of Cl_2:DOC ratios investigated; the only DBP for which this statement did not apply was monochloroacetic acid (MCAA), which was present at very low concentrations. Both the simplicity of the functional relationships and the overlapping of curves characterizing Cl_2 consumption and DBP formation for a wide range of experimental conditions demonstrate the utility of $|\Delta A_{272}|$ as a master parameter to monitor the progress of the chlorination process.

When the pH was adjusted over the range from 6 to 9, the same temporal patterns were observed in the parameters described above. However, the curves for Cl_2 consumption and DBP formation vs. $|\Delta A_{272}|$ in the systems at different pH's did not collapse to a single, coherent curve. Rather, as has been found in a number of prior studies, chloroform formation increase with increasing pH. In this study, the same trend applied to formation of $CHCl_2Br$ and CH, but formation of TCAA declined with increasing pH, and that of MCAA and DCAA was approximately independent of pH.

Chlorination of LW water after addition of up to 2.0 mg/L Br^- caused the overall Cl_2 consumption to increase slightly, with most of the increase occurring almost instantly. This increase can be attributed to a combination of the oxidation of Br^- to HOBr by Cl_2 and the rapid reaction of HOBr with some NOM sites.

As would be expected, increasing concentrations of Br^- in the solution prior to chlorination shifted the DBP population toward species containing increasing amounts of bromine. Other than this shift, the trends described above applied once again. The total molar concentration of trihalomethanes (THMs) increased with increasing Br^- concentration, while that of haloacetic acids (HAAs) did not change significantly. The distribution of organic bromine among THMs, HAAs, and haloacetonitriles (HANs) was approximately constant over time and over the range of Br^- concentrations investigated.

Decreasing solution temperature from 25°C to 10°C or 4°C led to a decline in Cl_2 consumption, and slight, inconsistent changes in DBP formation and $|\Delta A_{272}|$ development. Perhaps the key result from the tests at different temperatures is that a strong and similar relationship between DBP formation and $|\Delta A_{272}|$ existed at all three temperatures tested for all the DBPs analyzed. Thus, in these experiments, using $|\Delta A_{272}|$ as a master variable to indicate the progress of the reaction was, in a sense, even more reliable than using reaction time in that way.

Short-Term Kinetics of Chlorine Reactions With Lake Washington Water

In the final set of experiments with LW water, the reactions between NOM and free chlorine at reaction times from 0.05 s to 100 s were explored using the stopped-flow and quench-flow instrumentation. These experiments indicated that the reaction characteristics over this time frame are remarkably similar to those over longer time frames in more conventional kinetics experiments. For instance, the characteristic spectrum of $|\Delta A_{\lambda}|$ vs. λ obtained previously for reaction periods of several minutes to several days could be observed in the stopped-flow experiments after only 0.1 s of reaction. Furthermore, even for the shortest reaction times investigated, some DBPs were detected, plots of DBP concentration vs. time had an initial rise and were curvilinear thereafter, and DBP vs. $|\Delta A_{272}|$ plots followed the characteristic patterns seen in longer-term experiments.

As a result of the rapidity of the NOM reactions with chlorine, the hope that the very short-term studies would elucidate the initial steps in the reaction and allow the intermediates formed by those steps to be studied was not realized. On the other hand, the remarkable similarity between the short- and long-term results indicates that, as a practical matter, the key features of the reactions can be understood from essentially immediately after they are initiated to all future times based on conventional experiments.

MODELING DBP FORMATION

Conventional Kinetic Modeling of DBP Formation

The data from the experiments with Lake Washington water were used to develop and calibrate both a conventional kinetic model for DBP formation (describing DBPs as a function of reaction time and solution composition) and a model representing DBP formation as a function of $|\Delta A_{272}|$. These models were then applied both to modified LW water and to other chlorinated waters, to assess how widely the model parameters varied as a function of water quality. The conventional kinetic modeling was restricted to the THMs and HAAs, but the modeling that used $|\Delta A_{272}|$ as the independent variable included several other, trace DBPs.

The model construct for the conventional representation of the reaction kinetics (concentration vs. time) comprised a group of sequential and branching chlorination steps, each of which was assumed to be first-order with respect to the organic reactant and the halogen-based oxidant (either free chlorine or free bromine), hence second-order overall. The pathway for forming DBPs containing chlorine but no bromine involved three sequential chlorine incorporation steps, after each of which the chlorinated intermediate was assumed to be capable of reacting in two or three different ways. For singly and doubly chlorinated intermediates, the available paths were addition of yet another chlorine or cleavage of MCAA and DCAA, respectively. For the triply chlorinated intermediate, the paths included cleavage of TCAA or $CHCl_3$, or further reaction to form an unidentified product (which might, in turn, serve as an intermediate for formation of other DBPs, such as chloral hydrate).

To model the data for chlorination of Lake Washington water adequately, it was necessary to postulate the existence of two kinetically distinct reactive sites in the original NOM molecules. Both types of sites were presumed to follow the pathways described above, but the reaction rate constants applicable to the two sites were different. For simplicity, it was assumed that the ratio of the rate constants for the two sites was identical for each step in the reaction pathway.

All the data for THM and HAA formation at Cl_2:DOC ratios from 0.2 to 3.0 mg/mg could be modeled reasonably well using this approach. The best-fit parameters suggested that NOM molecules contain, on average, approximately 0.012 mole of precursor sites per mole of organic carbon, that 28% of those sites are in the group defined as rapid-reaction sites, that the rate constants for the rapid-reaction sites are 340 times as large as those for the slower-reacting sites, and that the first chlorine incorporation step is rate-limiting for formation of all the DBPs modeled.

The model under-predicted DBP formation at a Cl_2:DOC ratio of 0.1 mg/mg, because the model does not account for the virtually instantaneous formation of small amounts of DBPs. This deviation between the model and experimental results could be largely eliminated by postulating the existence of a small number of a third type of precursor site in the NOM molecules. However, that was not done in the current work, since those sites appear to make a significant contribution to DBP formation only under conditions that would rarely be of interest in practical systems.

The formation of $CHCl_3$ and HAAs in chlorinated water from several other sources was modeled using the same approach as described above. For this modeling, it was assumed that the fraction of the sites that could react rapidly and the ratio of the rate constants for the fast- and slow-reacting sites were identical to the values found for chlorination of LW water (28% and 340, respectively). In all cases, the experimental data could be fit fairly well using the model equations. The absolute values of the rate constants had to be adjusted slightly from one water to the next to obtain the best fit, but the range of values for a given rate constant fell within a factor of 3 for all the waters tested. In all cases, the modeling indicated that the first chlorination step was rate-controlling.

The formation of brominated DBPs was also modeled, using the data base for chlorination of LW water to which various concentrations of Br^- had been added. In general, the same model construct was used in this modeling as in modeling waters where the Br^- concentration was negligible. The oxidation of Br^- to free bromine was assumed to be quantitative and instantaneous upon addition of Cl_2, so that both free Cl_2 and free Br_2 species could act as halogenating and oxidizing agents.

Because of the complexity of the potential reaction pathways when both free Cl_2 and free Br_2 species are present, some additional assumptions had to be made to keep the number of adjustable parameters manageable. In particular, the ratio of the rate constants for bromination vs. chlorination (the branching ratio, *R*) was assumed to be the same for all precursors and intermediates. Other assumptions are detailed in the full report. The best-fit value of *R* was approximately 12, which is in reasonable agreement with values suggested by previous researchers. Also, because trace amounts of several DBPs formed almost instantaneously in these experiments, it was necessary to invoke the presence of a small concentration of extremely fast-reacting sites to model the data well. Even when that was done, however, the model tended to over-predict the impact of changes in the Br^- concentration on the formation of other DBPs. This over-prediction probably results from the assumption that the pool of precursor sites available for chlorination is identical to that available for bromination. To fit the data better, it appears necessary to postulate that a small population of precursor sites is available for bromination but not chlorination; no attempt was made to adjust the model in that way in the current work.

DBP formation in waters from sources other than Lake Washington that contained significant Br^- concentrations were modeled using the same construct. Again, several of the rate constants had to be altered to fit the data reasonably well, but in all cases a reasonable fit could be achieved. The best-fit branching ratios for these waters ranged from 2 to 20.

Modeling DBP Formation Using ΔA_{272} as a Master Parameter

While the mechanistic kinetic model described above conveys information about the rate of formation of DBPs over time, it requires the evaluation of several, source-water-specific fitting parameters that might change if the water quality changes. On the other hand, relationships between DBP formation and the decline in ultraviolet (UV) absorbance at 272 nm have been shown to be extremely robust and reliable over a broad range of halogenation conditions, even for waters of vastly different quality. In addition, the relationship between DBP formation and $|\Delta A_{272}|$ can be used as a secondary test of the validity of a proposed reaction scheme. That is, even if the scheme reproduces DBP vs. time observations quite well, if it does not also reproduce the relationship between DBP concentrations and $|\Delta A_{272}|$, it cannot be completely accurate. Finally, the ease with which $|\Delta A_{272}|$ can be measured means that, if it is as good an indicator of DBP formation as has been suggested, it could be used to estimate DBP formation in real time, thereby greatly reducing the need for sophisticated, conventional kinetic models. For all these reasons, an effort was made to evaluate whether the DBP data from chlorination of LW water could be modeled using $|\Delta A_{272}|$ as the sole independent parameter. Data for formation of chloroketones, acetonitriles, and chloral hydrate were considered, as well as those for $CHCl_3$ and HAAs.

The experimental data for most of the DBPs could be fit very well by such a model, using the assumptions that the unaltered sites and the first intermediate species (presumably a mono-chlorinated, aromatic compound) absorb UV light, but that intermediates and products formed by subsequent chlorination reactions do not. The estimated molar absorbance of the unaltered, reactive sites in NOM is in the same range as that of highly activated pure aromatic compounds, with the chlorinated intermediate being somewhat less UV-absorbing than the unaltered sites. Consistent with the analysis based on conventional kinetics, the $|\Delta A_{272}|$ analysis indicates that formation of the first halogenated intermediate is the rate-limiting step for DBP formation; it also suggests that the breakdown of the mono-, di- and trichlorinated intermediates becomes progressively easier in that order.

The fit of the model to the data for formation of chloral hydrate (CH) was less good than that to other stable DBPs, suggesting that formation of CH does not form from the same immediate precursor as $CHCl_3$ and TCAA. Additional modeling of this relationship is likely to provide insight into the pathway by which CH forms. The other DBPs for which the model did not reproduce the experimental data were those which clearly underwent decay reactions after they formed. Further development of the model would likely lead to insights into those processes as well. Additional developmental work is also needed to extend the model to the formation of brominated DBPs and to DBPs that undergo decay reactions after they form, and also to improve the coherence between the model and conventional kinetic modeling.

CONCLUSIONS AND RECOMMENDATIONS

In this project, mechanistic models for the formation kinetics of THMs and HAAs were developed in a manner that is consistent with the current understanding of DBP formation chemistry. The project demonstrated that modeling such reactions in a consistent and scientifically justifiable framework is, arguably, no more difficult, and can be accomplished using no more fitting parameters, than purely statistical modeling. Since the mechanistic modeling can provide insights into the chemical nature of the reactions and has the potential to reduce the amount of site-specific testing that would be required to model DBP formation under new conditions, it is preferred over statistical modeling in most cases. Modeling of DBP formation as a function of $|\Delta A_{272}|$ provides complementary information to classical kinetic modeling and also has practical advantages in terms of its ease of use and wide applicability.

Future research should focus on further development of mechanistic approaches to modeling DBP formation. In particular, effort should be devoted to further elucidation of the effects of pH and temperature on the values of the rate constants, and to enhancing the conceptual basis of the model so that it can explain both kinetic and spectroscopic observations. It is also necessary to apply the mechanistic model to a wider variety of water qualities, especially treated waters, so that the effects of water treatment can be ascertained. The integration of such data with the mechanistic model for a representative range of water qualities and reaction conditions will constitute an important advance in the evolution of drinking water chemistry and its engineering applications.

CHAPTER 1
INTRODUCTION

Chlorination of drinking water has been used to control waterborne infectious disease for more than a century and is widely recognized as one of the most effective public health measures ever undertaken (Raitt 2000). However, the formation of halogen-containing disinfection by-products (DBPs) via reactions of chlorine with natural organic matter (NOM) represents a public health risk and therefore places constraints on chlorination practice. Nevertheless, the practice of water chlorination is expected to continue and possibly expand in the future, in light of the degradation of the microbiological and chemical quality of many water sources, increasing worldwide scarcity of high quality potable water sources, and the need to ensure water quality and safety in emergency situations. Given this reality, a key issue associated with chlorination practice is how to predict the formation of DBPs in drinking water, so that exposures to those species can be quantified, health risks ascertained, and appropriate regulatory actions taken (Arbuckle et al. 2002; Simmons et al. 2002).

This report describes research directed at developing reliable and rational rate expressions for the formation of chlorinated DBPs in potable water. The vast majority of existing models for DBP formation are statistical rather than mechanistic, a feature that almost guarantees that they will be highly site-specific and not reliable for predictions outside the range of existing observations. To improve upon those models, this research focused on development, calibration, and testing of a model for DBP formation based on the stepwise sequence of reactions that constitutes the current qualitative understanding of the process. The goal was to include in the model explicit equations for (1) the rate of disappearance of precursor sites and of active chlorine and bromine, (2) the rates of formation and disappearance of halogenated reaction intermediates, and (3) the rate of formation of trihalomethanes (THMs) and haloacetic acids (HAAs). The model framework was intended to be sufficiently generic to be extended to other DBPs and total organic halide (TOX) species at a future time.

An additional goal of the project was to extend the authors' prior work relating DBP formation to the change in the ultraviolet (UV) absorbance of the water when it is chlorinated, a quantity referred to as the differential absorbance, ΔA_λ:

$$\Delta A_\gamma \equiv A_{\lambda,t} - A_{\lambda,0} \tag{1.1}$$

where $A_{\lambda,t}$ and $A_{\lambda,0}$ are the absorbance of the sample at wavelength λ at some time t after chlorination and at time zero (prior to chlorination), respectively. That prior work had suggested that DBP concentrations could be expressed much more simply as a function of ΔA_λ than as a direct function of time. As a result, since ΔA_λ can be measured relatively easily, rapidly, and inexpensively online, developing a model for DBP formation in terms of ΔA_λ offers the potential to greatly reduce the reliance of utilities and regulators on conventional kinetic models when estimating the DBP concentration at a given location and time.

To carry out the research, numerous experiments were conducted with a baseline water collected from Lake Washington (LW), in Seattle, Wash. In these experiments, the pH, temperature, and bromide concentration of the water were adjusted over a predetermined range, after which the water was dosed with free chlorine (as sodium hypochlorite, NaOCl). The decline of the free chlorine concentration, the change in the absorbance spectrum of the sample, and the formation of various DBPs were then analyzed after reaction times up to seven days. A unique feature of this research was the collection of data at very short reaction times (from < 1 s to 100 s), using sophisticated stopped-flow and quench-flow analytical instrumentation. The kinetics of DBP formation in several other raw drinking water sources from around the United States were also investigated, although the range of experimental conditions was less extensive in these experiments.

The data from the experiments with LW water were used to develop and calibrate both a conventional kinetic model for DBP formation (describing DBP concentrations as a function of reaction time and solution composition) and a model representing DBP formation as a function of ΔA_{λ}. These models were then applied to the other waters, to assess how widely the model parameters varied as a function of water quality. Although data for formation of many trace DBPs is reported here in addition to the data for THMs and HAAs, the conventional modeling was restricted to those two groups of regulated DBPs; the model representing DBP concentrations as a function of ΔA_{λ} was applied to a wider range of DBPs.

This report is organized into seven chapters. Following this introductory chapter are chapters providing a more extensive literature review on DBP formation kinetics and modeling (Chapter 2) and describing the experimental procedures (Chapter 3). Chapters 4 through 6 present, respectively, the experimental results, the conventional (DBP vs. time) modeling of the data, and the modeling of the data using ΔA_{λ} as the primary independent variable. Finally, the conclusions of the study and suggestions for future work are presented in Chapter 7.

CHAPTER 2
REVIEW OF PRIOR STUDIES OF DBP SPECIATION AND FORMATION KINETICS

Since the original recognition of chloroform as a DBP in chlorinated drinking water in the 1970s (Marx 1974; Rook 1974; Dowty et al. 1975), a wide variety of volatile, semi-volatile and non-volatile (hydrophilic) DBPs have been identified (Krasner et al. 1989; Richardson, Simmons, and Rice 2002). These DBPs include compounds containing either or both chlorine and bromine, and recent studies suggest that iodine-containing DBPs can form as well (e.g., Heller-Grossman, Manka, and Limoni-Relis 1993; Huang and Yeh 1997; Wu and Chadik 1998; Chellam and Krasner 2001; Roberts, Singer, and Obolensky 2002; Zhang and Minear 2002). A list of several important DBPs is provided in Table 2.1.

Typically, the species listed in Table 2.1 account for 40% to 70% of the total organic halogen (TOX) concentration in the treated water (Figure 2.1), with the rest of the TOX presumably being incorporated into larger NOM-type molecules or smaller but as yet unidentified NOM cleavage products (Stevens, Moore, and Miltner 1989; Singer and Chang 1989; Krasner et al. 1989; Symons et al. 1998; Frimmel, Hesse, and Kleiser 2000).

Despite significant recent advances in the identification of individual DBPs and the factors that control their formation, predicting DBP formation remains extremely challenging, due to the complexity of DBP reaction pathways, the wide variety of DBP species that form, and the multiple physico-chemical parameters that affect their formation. The current understanding of DBP formation processes and approaches for modeling the kinetics of those processes are described next.

DBP FORMATION: YIELD AND KINETICS

In general, the concentration of DBPs tends to increase with the concentration of dissolved organic carbon (DOC) in the water, the chlorine dose, and reaction time. However, the yield of individual DBP species (mass of DBP formed per unit mass of NOM present) and the kinetics of their release are complex functions of a variety of water quality parameters.

Parameters Affecting DBP Formation

NOM Properties and Concentration

The dissolved NOM in potable water sources comprises several types of compounds, including amino acids, proteins, carbohydrates, amino sugars, and carboxylic acids, but the precursors for most DBPs are thought to be humic species (Larson and Weber 1994; Martin-Mousset et al. 1997; Croué, Korshin, and Benjamin 2000). The reaction sites in humic species are predominantly activated aromatic functional groups referred to collectively as polyhydroxyphenolic acid (PHA) moieties (Harrington et al. 1996; Croué, Korshin, and Benjamin 2000). Other functional groups such as esters and ketones can also contribute to the formation of DBPs (Leenheer, Wilson, and Malcolm 1987; McClellan et al. 2000), but their contribution is less than that of PHAs.

Table 2.1
An abridged list of identified halogen-containing DBPs

Chemical formula	Name	Occurrence
$CHCl_xBr_y$, x+y=3	Trihalomethanes (THMs)	Major
$CH_xCl_yBr_zCOOH$, x+y+z=3	Haloacetic acids (HAAs)	Major
$CH_xCl_yBr_zCN$, x+y+z=3	Haloacetonitriles (HANs)	Substantial
$CH_xCl_yBr_zCOCH_3$, x+y+z=3	Haloketones (HKs)	Substantial
$CCl_xBr_yNO_2$, x+y=3	Trihalonitromethanes (THNMs)	Substantial
Cl-CN, Br-CN	Cyanogen chloride or bromide	Substantial
CCl_3-CHO	Chloral hydrate (CH)	Substantial

Although PHA moieties are widely believed to be the major groups involved in DBP formation, the reactivity of those groups is not completely understood, because of both the diversity inherent in the pool of NOM molecules and the difficulty of establishing a chemically specific description of the chlorine incorporation process (Leenheer 1981; Malcolm 1990; Bruchet, Rousseau, and Mallevialle 1990; Chin, Aiken, and O'Loughlin 1994; Dycus et al. 1995; Croué, Korshin, and Benjamin 2000). Even for relatively simple, pure aromatic compounds (e.g, phenol, resorcinol, hydroxybenzoic acids), the incorporation of chlorine involves multi-step and often branching reactions (Rook 1977; Boyce and Hornig 1983; Tretyakova, Lebedev, and Petrrosyan 1994; Gonzalez, Olson, and Rebenne 1996; Rios et al. 2000); the reactions of NOM are bound to be at least as complex and varied. As a result, neither the molar concentration of chemical sites in NOM available for reactions with chlorine nor their distribution (in terms of structure or chemical properties) can be established.

These limitations have necessitated the use of surrogate parameters to evaluate NOM reactivity. Of those, DOC is one obvious choice; for similar chlorination conditions (Cl_2 dose, pH, temperature, etc.), higher DOC levels usually correlate with greater DBP formation, although the correlation is not necessarily strong if the data set includes NOM from sources with widely differing water qualities. Absorbance of ultraviolet light at 254 nm (A_{254}), is thought to be a good surrogate for the concentration of activated aromatic groups in PHA moieties (Traina, Novak, and Smeck, 1990; Wang, Pant, and Langford 1990; Novak, Mills, and Bertsch 1992; Korshin, Li, and Benjamin 1996; Croué, Korshin, and Benjamin 2000) and is usually better correlated than DOC with the yield of DBPs (Edzwald, Becker, and Wattier 1985; Edzwald 1993; Korshin, Li, and Benjamin 1997b; Huang and Yeh 1997; Chellam and Krasner 2001). A third parameter, computed as the ratio of A_{254} to DOC and known as the specific ultraviolet absorbance at 254 nm ($SUVA_{254}$), has also been widely used as a surrogate for NOM reactivity in DBP formation reactions. Higher yields of DBPs are almost invariably observed for NOM with higher $SUVA_{254}$ values (Korshin, Benjamin, and Sletten 1997; Croué, Korshin, and Benjamin, 2000; Karanfil, Schlautman, and Erdogan 2002; Kitis et al. 2001, 2002).

Solution pH

In a given water chlorinated under otherwise identical conditions, the yield of THMs invariably increases with increasing pH (Heller-Grossman, Manka, and Limoni-Relis 1993;

Krasner et al. 1989; Arora, LeChevallier, and Dixon 1997; Singer, Obolensky, and Greiner 1995; Miller and Uden 1983; Diehl et al. 2000). By contrast, the yields of HAAs and TOX usually decrease (Stevens et al. 1976; Miller and Uden 1983; Uden and Miller 1983; Carlson and Hardy 1998; Diehl et al. 2000), although these latter trends are less consistent than the trend in THMs. The effects of pH on the formation of many minor DBP species (e.g., CH, HKs, and HANs) are less clear—the initial yield of these species may increase with an increase in pH, but because many of them can undergo base-catalyzed hydrolysis, their concentrations sometimes decrease if the pH is raised higher, or if they are held at an alkaline pH for long periods.

Disinfectant (Oxidant) Concentration

The formation rate and ultimate concentrations of THMs and HAAs formed in a given water increase with increasing Cl_2 dose. At a fixed Cl_2 dose, THM production increases with increasing DOC concentration if a free chlorine residual is maintained, but decreases with increasing DOC concentration if the Cl_2 is completely depleted (Adin et al. 1991). This latter result suggests that, at low Cl_2-to-DOC (Cl_2:DOC) ratios, chlorinated intermediates form but are not necessarily converted to THMs.

The effects of Cl_2 and DOC concentration on HAA formation are more complex, because of the formation of mono- and di-halogenated HAAs along with tri-halogenated ones. As one would expect for a sequential reaction process, the HAA speciation shifts toward more highly halogenated forms as the Cl_2:DOC ratio increases. For instance, Miller and Uden (1983) showed that formation of dichloroacetic acid (DCAA) is favored over trichloroacetic acid (TCAA) at lower Cl_2:DOC ratios, but the preference is reversed at higher ratios.

Aside from the complexity of the DBP formation reactions themselves, the relationship between Cl_2 dose and DBP formation is confounded by the complexities of chlorine decay reactions. Chlorine dosed into a water supply can be consumed in at least three ways. Some chlorine oxidizes NOM sites and is reduced to Cl^-, some becomes incorporated into NOM molecules, and some might oxidize other sites or other molecules that have no relationship to DBP formation. The non-uniformity of organic functional groups in NOM, both those generating DBP species and those undergoing oxidation without chlorine incorporation, further complicates the situation. Due to these difficulties, most models of DBP formation kinetics use the initial Cl_2 dose to characterize the Cl_2 concentration throughout the reaction period, even though that concentration might change by an order of magnitude or more during the reaction.

Reaction Time

Changes in the DBP formation rate over time have been explored primarily for THMs (most frequently, chloroform), with less emphasis on formation of HAAs and TOX, and almost no work on other DBPs. Frequently, greater than 50% of the ultimate THM and HAA formation is reported to occur within a few minutes to tens of minutes after chlorine addition (e.g., Qualls and Johnson 1983; Westerhoff et al. 2000). Thereafter, DBP formation proceeds much more slowly, but nevertheless measurably, for at least several days. Most models focus on $t > 100$ minutes, and they often are reasonably accurate (± 30%) from that time forward. However, that result is not surprising, since the slow change in THM concentration can be well modeled without much mathematical sophistication, and the models tend to incorporate many fitting parameters. In cases where researchers have attempted to account for both short- and long-term DBP formation, they have often found it necessary to use different sets of coefficients or different equations for different exposure times (Urano, Wada, abd Takemasa 1983; Engelholm and Amy 1983; Morrow and Minear 1987; Westerhoff et al. 2000).

Virtually all models of DBP formation kinetics predict a monotonic increase in the concentrations of all DBPs over time. This prediction is usually accurate for THMs and HAAs, because they are generally stable in water, although they can degrade under certain specialized conditions (e.g., via photooxidation [Pehkonen, Siefert, and Hoffmann 1995], hydrolysis [El Din, Arain, and Ahammoud 1998], or reductive dehalogenation [Croué and Reckhow 1989]). Brominated HAAs and TCAA may be dehalogenated under strongly reducing conditions (e.g., in the absence of dissolved oxygen, at the surface of galvanized pipes [Korshin and Jensen 2001]). The degradation of DBPs might also be mediated by microorganisms (Arora, LeChevallier, and Dixon 1997; Castro, O'Shea, and Wang 1996) or occur via abiotic hydrolysis, which can cause a rapid disappearance of HANs, HKs, and THNMs at slightly elevated pH (Suffet, Brenner, and Silver 1976; Oliver 1983; Croué and Reckhow 1989; Korshin et al. 2002a,b).

Bromide Concentration

Bromide is rapidly oxidized to hypobromous acid (HOBr) by free chlorine (Kumar and Margerum 1987; Westerhoff et al. 1998), and the HOBr can then react with NOM in either substitution or oxidation reactions (Leung, Peters, and Perry 1982; Adin et al. 1991; Clark et al. 1996; Rathbun 1996a; Cowman and Singer 1996; Richardson et al. 1999). In most cases, HOBr substitutes into NOM much more rapidly, and oxidizes NOM more slowly, than hypochlorous acid (HOCl) or hypochlorite ion (OCl^-), so it can have a large effect on the speciation and yield of DBPs (Cowman and Singer 1996; Nokes, Fenton, and Randall 1999; Westerhoff et al. 2000).

Due to the rapidity of the Br incorporation reactions, and also because bromine incorporation into DBPs is thought to exacerbate their health effects, the concentration of bromide in drinking water is of special significance. Several semi-quantitative mechanistic models of bromine incorporation into HAAs and THMs have been developed (Heller-Grossman, Manka, and Limoni-Relis 1993; Cowman and Singer 1996; Petri et al. 1997; Nokes et al. 1999), along with numerous statistical (i.e., non-mechanistic) models (e.g., Urano et al. 1983; Engelholm and Amy 1983; Amy, Chadik, and Chowdhury 1987; Harrington, Chowdhury, and Owen 1992; Rathbun 1996a; Westerhoff et al. 2000). Some of these models are discussed in the following section of this chapter.

Models for the Kinetics of Chlorine Consumption and DBP Formation

The reactions leading to DBP formation consume free chlorine and reactive sites in NOM, and generate various halogenated products, including both small, identifiable DBPs and larger halogenated organics. Of these species, only the free oxidant and the concentrations of the identifiable DBPs are amenable to direct quantification, and of those only a small subset has actually been studied in any detail. As a result, kinetic models of DBP formation tend to focus on those species.

Models for the Kinetics of Chlorine Consumption

Although the equilibria and kinetics of reactions between dissolved halogen species and several pure compounds have been studied in detail (e.g., Kumar and Margerum 1987; Fogelman,

Table 2.2
Chlorine decay kinetic models (adapted from Powell et al. 2000)

Classification of model	Analytical expression for chlorine consumption	Eq. #	Notes
First order Cl_2 decay	$C = C_0 e^{-k_b t}$	(2.1)	Simplest and most widely used model
Limited first order	$C = C^* + (C_0 - C^*)e^{-k_b t}$	(2.2)	First order decay approaching a finite, minimum Cl_2 residual at long t
Parallel first order	$C = C_0 z e^{-k_{bfast} t} + C_0(1 - z)e^{-k_{bslow} t}$	(2.3)	Considers two, kinetically distinct Cl_2 consumption processes
First order with respect to Cl_2 and another reactant	$C = \frac{U - C_0}{\frac{U}{C_0}\exp[W(U - C_0)t - 1]}$	(2.4)	U and W relate to initial concentration of second reactant, reaction stoichiometry, and rate constant (see original article for details)
Second order Cl_2 decay	$C = \frac{C_0}{1 + C_0 k'_b t}$	(2.5)	Predicts steeper initial decline in Cl_2 concentration than first order model, as is often observed experimentally
nth order Cl_2 decay	$C = [k_b''' t(n - 1) + C_0^{-(n-1)}]^{-\frac{1}{n-1}}$	(2.6)	

Walker, and Margerum 1989; Wang and Margerum 1994; Wang et al. 1994; Beckwith, Wang, and Margerum 1996), the difficulty of characterizing processes that involve both chlorine (and/or bromine) species and NOM have impeded the development of chlorine consumption models that take the underlying reaction mechanisms into consideration. Powell et al. (2000) grouped models of chlorine consumption into the six categories shown in Table 2.2. They concluded that, although some of the more complicated models fit the available data better than a first-order decay model, the marginal improvement in fit to the data does not justify the added complexity.

The model equations shown in Table 2.2 have been used with a reasonable level of success only for relatively long-term chlorine decay kinetics. Even in recent models of chlorine decay and DBP formation that incorporate parallel "fast" and "slow" DBP formation reactions, the fast reactions are typically represented as occurring over periods of at least 4 to 6 hours (Jadas-Hecart et al. 1992; Clark et al 2001; Gang et al. 2002, 2003). More highly mechanistic models of halogen incorporation into NOM (e.g., McClellan et al. 2000, Gallard and von Gunten 2002) sometimes address DBP formation over much shorter periods, but the amount of data available for such modeling is relatively scant. As a result, to the extent that reactions occurring immediately following the injection of disinfectant into drinking water have been modeled at all, they have most often been represented as occurring instantaneously.

Models for the Kinetics of DBP Formation

Approaches for modeling DBP formation kinetics can be categorized into the following groups:

1. Statistical models that do not rely on presumptions about the mechanisms of chlorine incorporation into NOM and the ensuing DBP cleavage reactions, but rather seek to describe the experimental data using best-fit empirical equations.
2. Models based on *a priori* assumptions concerning the form of the integrated rate equations (e.g., first- or second-order kinetics). Mathematically, many of these models utilize equations with the same form as those shown in Table 2.2.
3. Mechanistic models based on the presumed chemistry of the chlorination reaction. Due to the complexity of NOM chlorination reactions, these models do not necessarily yield explicit analytical expressions for DBP formation kinetics, but they can be integrated numerically to generate predictions for DBP concentrations as a function of reaction time and initial reaction conditions. They also allow alternative hypotheses concerning the kinetics of the rate-limiting steps in the overall reaction sequence to be evaluated.
4. A model in which the depletion of reactive sites in NOM and the formation of DBPs or TOX are related to the differential absorbance, ΔA_λ, of the sample (Korshin at al. 2002a; Li, Korshin, and Benjamin 1998). This approach differs from those described above in that it does not yield predictions of DBP concentrations over time based on the conditions at time zero in the chlorinated water, but rather provides a technique for estimating the DBP concentrations based on measurements of ΔA_λ as the reaction occurs. These measurements can be carried out on-line, rapidly, and with high precision. Thus, in a sense, this approach obviates the need for predicting DBP concentrations, since it allows DBP concentrations to be estimated directly from data gathered in the reactive system in real time.

Some key features of these approaches are discussed below.

STATISTICAL MODELS FOR THE YIELD AND KINETICS OF DBP FORMATION

Statistical models produce regressions relating DBP concentrations to known reaction conditions at the beginning of the chlorination process (e.g., initial NOM and Cl_2 concentrations, $SUVA_{254}$, pH, etc). Most such models use the chlorine concentration in the original solution (i.e., based on the Cl_2 dose) to characterize the system throughout the reaction period, even though that concentration might change dramatically during the reaction, as suggested by the equations in Table 2.2. Although a few groups of researchers have tried to incorporate Cl_2 decay kinetics into equations for DBP formation (Peters, Young, and Perry 1980; Montgomery Watson 1993; Harrington, Chowdhury, and Owen 1992; Black, Harrington, and Singer 1996; Westerhoff et al. 2000), the generality of these equations has not yet been established. A few authors (Rodriguez, Serodes, and Morin 2000; Milot, Rodriguez, and Serodes 2000, 2002) have used a probabilistic, rather than deterministic approach in such modeling, i.e., their models predict the probability that a utility will produce water in which the DBP level exceeds a specified value. A representative group of statistical equations that have been proposed to model DBP formation kinetics is presented in Table 2.3.

The equations in Table 2.3 are shown as presented in the cited publications; readers are referred to those papers for additional details. In the Sung et al. (2000) paper, *a* and *b* are empirical fitting parameters, and in the Golfinopoulos and Arhonditsis (2002) paper, the

Table 2.3
Selected statistically-based equations to model DBP formation

Equation	Eq. #	Reference
$[THM] = 0.0035\left[\exp\left(-\frac{4470}{T}\right)\right](pH - 2.8)[DOC][Cl_2]_0^{0.25}\ t^{0.36}$	(2.7)	Urano, Wada, and Takemasa (1983)
$[THM] = 0.0031(UV_{254} \cdot [DOC])^{0.44}[Cl_2]_0^{0.409}\ t^{0.265}T^{1.06}(pH - 2.6)^{0.715}([Br] + 1)^{0.0358}$	(2.8)	Amy, Chadik, and Chowdhury (1987)
$[TTHM] = 0.00309[(DOC)(DUV - 254)]^{0.440}(Cl_2)^{0.409}t^{0.265}T^{1.06}(pH - 2.6)^{0.715}(Br + 1)^{0.036}$	(2.9)	Harrington, Chowdhury, and Owen (1992)
$[THM] = 7.1(UV_{254})^{0.534}[DOC]^{0.004}([Cl_2] - 7.6[NH_3])^{0.224}t^{0.225}T^{0.48}(pH - 2.6)^{0.719}([Br] + 1)^{2.01}$	(2.10)	Black, Harrington, and Singer (1996)
$[THM] = 14.6[(UV_{254})^{0.849}[Cl_2]^{0.206}t^{0.306}(pH - 3.8)^{1.01}]$	(2.11)	Rathburn (1996a,b)
$TTHM = 23.9[(TOC)(UVA)]^{0.403}(Cl_2)^{0.225}(Br)^{0.141}1.027^{T-20}1.156^{(pH-7.5)}t^{0.264}$	(2.12)	Solarik et al. (2000)
$[TTHM] = a[OH^-]^j([Cl_2](1 - \exp(-kt))^m)(UV_{254})^n(algae)^p$	(2.13)	Sung et al. (2000)
$[HAA5] = b[OH^-]^i([Cl_2](1 - \exp(-kt))^k)(UV_{254})^l$		
$[TTHM] = 13.54\ln[Cl_2] - 14.47pH + 230.3[Br] - 139.6[Br]^2 - -25.3S + 110.6Sp - 6.59(T \cdot Sp) + 1.48(T \cdot [Cl_2])$	(2.14)	Golfinopoulos and Arhonditsis (2002)

parameters S and Sp account for seasonality (they stand for summer and spring, respectively), with quantitative definitions given in the original paper.

Sophisticated models and computer programs have been developed to simulate DBP formation in practical applications based on equations such as those in Table 2.3 (Harrington, Chowdhury, and Owen 1992; Westerhoff et al. 2000). While those models often fit the data for the systems used in the calibration step, evaluating the fitting parameters in other systems requires an extensive study of DBP formation in those systems, limiting the value of the models as true predictive tools. Additionally, as noted above, even in the systems for which they are calibrated, these models are usually reliable only for reaction times that exceed by several orders of magnitude the times presumed to be associated with the elementary chlorine incorporation steps (milliseconds to seconds) (Christman et al. 1989; Tapudurti and Haas 1991; Rebenne, Gonzalez, and Olson 1996; Pomes et al. 1999; Chaidou et al. 1999; Gonzalez, Olson, and Rebenne 1996; Rios et al. 2000; Gallard and von Gunten 2002). At least in principle, this limitation can be circumvented by using formal kinetic models to describe DBP formation.

MECHANISTIC KINETIC MODELS FOR PREDICTING DBP FORMATION

In comparison with the diversity of statistical equations that have been proposed to model DBP formation, the functional forms of mechanistic models are more limited. These models often incorporate *a priori* assumptions of first- or second-order kinetics, so that the integrated rate equations have forms similar to those shown for chlorine consumption in Table 2.2. Such is the case, for example, for the models presented by El Dib and Ali (1995), Casey and Chua (1997), Ivancev-Tumbas et al. (1999), El-Shafy and Grunwald (2000), and Gang, Clevenger, and Banerji (2003).

In some cases, the yield of individual DBPs has been modeled as being linearly proportional to the chlorine consumption. For example, Gang et al. (2002, 2003) found that the yields of the four chlorinated and brominated THMs (THM4) ranged from 31 to 48 μg per mg Cl_2 consumed, and those for the nine corresponding HAAs (HAA9) were between 16 and 28 μg per mg Cl_2 consumed. Based on the presumption of first-order kinetics of chlorine consumption by each of two kinetically dissimilar sites (slow and fast, similar to a concept proposed by Qualls and Johnson [1983]), Gang et al. (2002, 2003) derived the following kinetic expressions:

$$C(t) = C_0[fe^{-k_R t} + (1-f)e^{-k_S t}] \tag{2.15}$$

$$THM4 = \alpha C_0[1 - fe^{-k_R t} - (1-f)e^{-k_S t}] \tag{2.16}$$

$$HAA9 = \beta C_0[1 - fe^{-k_R t} - (1-f)e^{-k_S t}] \tag{2.17}$$

where $C(t)$ and C_0 are the free chlorine concentrations at time t and time 0, respectively, k_R and k_S are the kinetic constants for the rapid and slow chlorine consumption reactions, respectively, α and β are the yields of THM4 and HAA9, respectively, and f is the fraction of chlorine consumption attributed to rapid reactions. For Mississippi River NOM fractions with molecular weights ranging from ca. 500 to >10,000, the ratio of the kinetic constants for the rapid and slow chlorine consumption reactions was between 90 and 145, and the rapid reactions accounted for 28% to 34% of the overall chlorine consumption (i.e., the value of f was 0.28 to 0.34).

Gallard and von Gunten (2002) also represented NOM molecules as having two kinetically distinct sites for reactions with chlorine in their model of long-term chlorine demand and THM formation. They reported second-order overall reaction rate constants for the slow-reacting sites (first-order each with respect to Cl_2 and NOM sites) between 0.01 and 0.03 $M^{-1}s^{-1}$ in the pH range 7-9 for surface water NOM; the rate constant for a single groundwater was, unexpectedly, much larger (0.124 $M^{-1}s^{-1}$). The fast-reacting sites accounted for 15-30% of the total pool. The researchers hypothesized that those sites were similar to reactive sites in resorcinol or other β-diketone or β-ketoacid structures, while the slow reaction sites were less strongly activated phenolic groups.

In other studies, analytical expressions for THMs and HAAs have been generated based on some assumed reaction rate expression in combination with statistical evaluation of the fitting parameters. For instance, Clark et al. (2001) developed the following expressions to characterize chlorine consumption and DBP formation, assuming that the reaction rate was first-order in both Cl_2 concentration and the concentration of precursors, but that the reaction stoichiometry was not necessarily 1:1 between the two reactants:

Expression for Cl_2 consumption:
$$C_A = \frac{C_{A0}(1-K)}{1-Ke^{-ut}} \tag{2.18}$$

Expression for DBP release:
$$\text{DBP}_i = A_i(C_{A0} - C_A) \tag{2.19}$$

In these expression, C_A is the concentration of chlorine at time t, C_{A0} is the initial concentration of chlorine, A_i is the yield of DBP species i per unit amount of chlorine consumed (μM per mg), and K and u are composite constants that depend on the initial concentrations of the reactants, the reaction stoichiometry, and the reaction rate constant. Although the above equations can be derived from the model assumptions, and the values of A_i, K, and u have some theoretical interpretation within that context, those values must be determined from empirical data and are, effectively, fitting parameters. Clark et al. (2001) presented the following expressions to correlate those parameters with water quality data:

$$u \equiv M(1-K) \tag{2.20}$$

$$M = e^{3.96}e^{-0.305(\text{Br}^-)}e^{0.0145(\text{Cl}_0 \cdot \text{pH})}(\text{Cl}_0)^{-2.32}e^{8.46P}e^{-0.231\,\text{pH}} \tag{2.21}$$

$$K = e^{1.89}(\text{pH})^{-0.13}(\text{Br}^- + 1)^{0.10}(\text{Cl}_0)^{-0.75} \tag{2.22}$$

where P is the molar ratio of the bromide concentration to the sum of the bromide and chloride concentrations.

This combined kinetic/statistical model worked reasonably well for 12 DBPs ($R^2 > 0.9$), but not for MCAA, whose concentration was deemed to be too low for an appropriate level of quantification. The equations predict that higher levels of bromide favor the formation of more highly brominated DBPs and accelerate their release. However, the shortest time for which the experimental data were generated was 6 hours, the chlorine dose was much higher than is typical in water treatment applications (11 to 25 mg/L, for a water containing 2.9 mg/L DOC), and the bromide levels investigated ranged up to 4.5 mg/L, as compared with < 0.4 mg/L in typical drinking water sources.

The formation of THMs under more realistic conditions (in four treated waters in Paris) was explored by Westerhoff et al. (2000). In this study, the authors used a model that incorporated two, parallel (fast and slow) first-order reactions to represent Cl_2 decay. The expression for DBP formation included a term that was proportional to the Cl_2 consumption, along with other terms related to the initial DOC and bromide concentrations, the temperature and pH of the water, and the reaction time. Statistical expressions were used to relate the rate constant for Cl_2 decay and the other empirical constants in the model to water quality parameters.

A good correlation between the observed and predicted levels of THMs in the treated water was reported. However, the good performance of this and similar kinetic models is achieved via elaborate combinations of *a priori* assumptions, careful statistical fitting, and application of various formalisms to describe chlorine consumption and DBP release at short and long reaction times. The simplicity of the first- and second-order rate expressions used in these models is thus partially counterbalanced by the need to embed complex fitting formulations for the rate constants.

Presumably, the limitations described above could be overcome if the reaction mechanisms for chlorine incorporation were reflected in the kinetic expressions. As discussed previously, the complexity of halogenation pathways, the variety of reaction sites in NOM, and presence of numerous chlorine consumption mechanisms combine to make a consistent and systematic realization of this approach very difficult. Nonetheless, there are essential elements of such an approach that, even when utilized only partially, make it quite distinct from the other approaches described above.

Specifically, a key element in any fundamentally sound model must be the recognition of the step-wise incorporation reactions of chlorine, leading to formation of intermediates from which individual DBPs subsequently cleave. Indeed, virtually all DBP species have been shown to form via sequences of reactions that are initiated by attack of chlorine on two main types of sites: activated aromatic units associated with the PHA moiety, and less prevalent ketone groups (Larson and Weber 1994; Tretyakova, Lebedev, and Petrrosyan 1994). The subsequent steps leading to formation of identifiable DBPs are not well understood except in a few cases, with the formation of THMs via the classic haloform reaction being the prominent example (Larson and Weber 1994). In that sequence, a ketone site undergoes three halogenation steps to form chloroform, as shown schematically in Figure 2.2.

Chlorine interactions with aromatic sites are more complex than those shown in Figure 2.2. Virtually all mechanistic models of THM formation from aromatic precursor sites postulate that the reaction sequence starts with multiple attacks by chlorine on the ring, generating a cyclic but non-aromatic chlorinated compound, such as shown schematically in Figure 2.3 (Rook 1977; Boyce and Hornig 1983; Tretyakova, Lebedev, and Petrrosyan 1994). This intermediate is then envisioned to undergo a series of transformations that open the ring, incorporate more halogens into the products, and eventually cleave the molecule into smaller products. Some of the products are one- and two-carbon molecules (THMs and HAAs), but others are larger halogenated fragments that contribute to the unidentified fraction of the TOX.

Recent studies (e.g., Rios et al. 2000) indicate that mechanisms other than that shown in Figure 2.3 might be important in the chlorination of activated aromatic compounds. However, the assumption that the formation of tri-halogenated species such as THMs always involves at least three sequential chlorine incorporation steps remains undisputed.

Analytic solutions to the differential equations describing the kinetics of these reactions, even for a single type of site at varying chlorine dosages, are generally unavailable, but numerical solutions can be developed. Alternatively, simplified equations can be produced for limited sets of conditions (e.g., very high chlorine concentrations, or kinetic control by a specified step [Jimenez et al. 1993]).

Adin et al. (1991) proposed a model for THM formation in which two halogenated intermediates are formed prior to the release of THMs. The authors did not develop a mechanistic model for chlorine consumption, choosing to use a statistical model instead. The resulting expression for the rate of THM formation at high concentrations of chlorine had the following form:

$$[THM] = A_0 K_1 K_2 \left[\frac{1}{(K_1+K_3)(K_2+K_4)} + \frac{\frac{1}{(K_1+K_3)}\exp(-(K_1+K_3)t) - \frac{1}{K_2+K_4}\exp(-(K_2+K_4)t)}{K_1+K_3-(K_2-K_4)} \right] \quad (2.23)$$

The use of models that account for the formation of halogenated intermediates has been prevalent in modeling the formation of DBPs containing both chlorine and bromine. For example, Cowman and Singer (1996) described a speciation model for HAAs based on the assumptions that all HAAs containing a given number of halogen atoms derive from the same precursor molecule and that the relative amounts of substitution with bromine and chlorine are fixed by the HOBr/HOCl concentration ratio and the (fixed) ratio of reactivities of these two oxidants. It was also assumed that the conversion of the intermediates to HAAs was not the rate-limiting step. Using an HOBr/HOCl reactivity ratio of 10, the authors reported a reasonably good fit between the model output and the HAA speciation resulting from chlorination of humic acid in the presence of various concentrations of Br^-.

Using a similar approach, Nokes, Fenton, and Randall (1999) examined the speciation of THMs at varying Br^-/Cl_2 concentration ratios. They assumed that the trihalogenated precursors of THMs are formed by three successive steps, each of which substitutes either bromine or chlorine into an activated site in the NOM. This leads to six pairs of competing bromination/chlorination reactions, whose rate constant ratios (denoted as k_b/k_c and referred to as branching ratios) were estimated using THM data from the chlorination of 17 waters from New Zealand. The k_b/k_c branching ratios for the different reactions were determined to range from 4 to 15. From a plot of the ratio of total bromine to total chlorine present in the THMs as a function of the Br^-/Cl_2 ratio in the water immediately after chlorination, an apparent overall k_b/k_c ratio of 9.1 was obtained. This ratio is close to that estimated by Cowman and Singer (1996).

A more detailed mechanistic study of DBP formation kinetics was presented by McClellan et al. (2000). This model is based on the hypothesis that NOM contains three classes of kinetically distinct reactive functional groups—one group that reacts with chlorine virtually instantaneously (relative to the time scale of the researchers' experiments, which was from 6 minutes to 7 days); another group (denoted S_1) that reacts with chlorine less rapidly and for which the rate-limiting step is first-order in both NOM and chlorine; and a third group (S_2), for which the rate-determining step is first-order in NOM only, leading to an active form of S_2 that reacts rapidly with HOCl in a second-order reaction. Site S_2 was hypothesized to correspond to ketones that undergo the classical haloform reaction with halogens (shown in Figure 2.2). For sites S_1 and S_2, a series of fast steps associated with chlorine consumption was presumed to follow the rate-limiting step, leading to a variety of halogenated and oxidized organic species, carbon dioxide and chloride. The conceptual model underlying this approach is shown in Figure 2.4.

Sites S_1 and S_2 were hypothesized to have different stoichiometries in their reactions with chlorine, and chloroform, DCAA and TCAA were assumed to cleave from the same intermediates (the formation of MCAA was not taken into account). No distinction was made among intermediates with varying degrees of chlorination with respect to their tendency to release DBPs. Based on this conceptual model, the following rate expressions were obtained for the formation of the DBPs:

$$r_{THM4} = \alpha_{1,\,THM} k'_1 [S_1][Cl_2] + \alpha_{2,\,THM} k'_{22} [S_2][HOCl] \qquad (2.24)$$

$$r_{DCAA} = \alpha_{1,\,DCAA} k'_1 [S_1][Cl_2] + \alpha_{2,\,DCAA} k'_{22} [S_2][HOCl] \qquad (2.25)$$

$$r_{TCAA} = \alpha_{1,\,TCAA} k'_1 [S_1][Cl_2] + \alpha_{2,\,TCAA} k'_{22} [S_2][Cl_2] \qquad (2.26)$$

In these equations, k_i values are apparent kinetic rate constants, and $\alpha_{i,DBP}$ is the molar yield of the DBP per mole of sites consumed. The apparent rate constants were related to the intrinsic constants using expressions reflecting the pH-dependence of the speciation of chlorine and a pseudo-Arrhenius expression to account for temperature effects:

$$k'_1 = (k_{11}\alpha_{HOCl} + k_{12}(1 - \alpha_{HOCl}))\theta_1^{T - T_{ref}} \qquad (2.27)$$

$$k'_{22} = k_{22}\theta_{22}^{T - T_{ref}} \qquad (2.28)$$

where θ is an empirical constant related to the activation energy for the reaction.

The mechanisms that cause individual DBPs to cleave from their precursors were hypothesized to include uncatalyzed hydrolysis, base-catalyzed hydrolysis, and oxidation with HOCl. Provision was made to incorporate these alternative mechanisms into the formal kinetic model. Overall, the mechanistic model employed 21 adjustable parameters. The data used to calibrate the model were obtained from experiments using one water source (Lake Gaillard, [in Connecticut]), in which the DOC was varied from 2.1 to 1.1 mg/L by dilution, and the pH was varied from 6.3 to 9.0. The temperature ranged from 2 to 25°C, the Cl_2 dose from 1.3 to 4.0 mg/L, and the reaction time from 6 minutes to 172 hours.

The R^2 values for the global fit that included the data points for THMs, DCAA and TCAA, were 0.92, 0.88 and 0.85 for the mechanistic model and for two alternative approaches that used statistical functions or second-order kinetic expressions, respectively. For another water quality (Central Arizona Project), the mechanistic model performed comparably to the others, but none of the models performed very well.

THE ΔA MODEL FOR CORRELATING SPECTRAL DATA WITH DBP FORMATION

Strong correlations between DBP formation potentials and $SUVA_{254}$ have frequently been reported in the literature (e.g., Korshin, Benjamin, and Sletten 1997; Croué, Korshin, and Benjamin 2000; Karanfil, Schlautman, and Erdogan 2002; Kitis et al. 2001). In addition, absorbance of light at 254 nm (without normalization by DOC) has been incorporated into most of the statistical equations developed to predict the kinetics of DBP formation, as indicated in

Table 2.3. These choices reflect the widely accepted belief that aromatic moieties in NOM are overwhelmingly responsible for both its light properties and its tendency to incorporate halogen and form DBPs.

As noted previously, almost all of the above-mentioned correlations incorporate the absorbance of light prior to halogenation as a parameter, but they do not explicitly account for the change in absorbance that inevitably accompanies the reactions. On the other hand, a few researchers have focused on the change in absorbance as a critical and highly informative parameter to characterize the reactions of interest. For instance, Jimenez, Dominguez, and Silverio (1993) observed that the absorbance of light at 257 nm steadily diminished as the reaction progressed and utilized kinetic profiles of A_{257} vs. time to estimate rates of THM formation at pHs between 7 and 9. The data were interpreted based on a simple mechanistic model that presumed the formation of a single halogenated intermediate.

A more extensive set of investigations of the relationship between changes in absorbance (sometimes referred to as differential absorption spectroscopy) and NOM halogenation has been conducted by Korshin and co-workers (Korshin, Li, and Benjamin 1997a, 1999; Li, Benjamin, and Korshin 2000; Korshin et al. 2002a,b). Those researchers quantified the change in absorbance (the differential absorbance, ΔA_λ) with very high precision at a selected wavelength (typically 272 nm) and correlated ΔA_{272} with the formation of both TOX and several individual DBPs. Formally, the differential absorbance is defined as follows:

$$\Delta A_\lambda \equiv A_{\lambda,\,t} - A_{\lambda,\,0} \tag{2.29}$$

where $A_{\lambda,t}$ and $A_{\lambda,0}$ are the absorbance of the sample at wavelength λ at some time t after chlorination and at time zero (prior to chlorination), respectively.

Because the absorbance at 272 nm (and, in fact, throughout the UV region of the spectrum) always decreases upon chlorination, ΔA_{272} is always negative. With increasing chlorination dose or contact time, the decline in A_{272} increases, i.e., ΔA_{272} becomes progressively more negative. To avoid the awkwardness of describing ΔA_{272} as "increasing in a negative direction," throughout the report, trends are presented as a function of the absolute value of A_{272}, i.e., as a function of $|\Delta A_{272}|$.

Korshin, Li, and Benjamin (1997a) found that TOX was strongly and linearly correlated with $|\Delta A_{272}|$. The correlation passed through the origin, and its slope was remarkably constant over a wide range of pHs, halogen doses, reaction times and NOM sources. Although additional studies indicated that the slope of the correlation was somewhat different for NOM fractions with either extremely high or extremely low aromaticities, the contributions of such fractions to the DOC of natural waters are typically low.

The researchers also identified correlations between $|\Delta A_{272}|$ and the formation of individual DBPs, notably THMs and HAAs (Korshin, Li, and Benjamin 1999; Li, Benjamin, and Korshin 2000; Korshin et al. 2002a,b). In contrast to the correlations for TOX, the relationships of THM and HAA concentrations with $|\Delta A_{272}|$ were not linear, they were affected by the solution pH, and, most significantly, the release of individual DBPs became noticeable only for $|\Delta A_{272}|$ values exceeding a certain threshold value. This latter observation was interpreted as indicating that, at $|\Delta A_{272}|$ values below the threshold, halogen was being incorporated into NOM molecules to form TOX, but those molecules were not yet breaking down to release identifiable DBPs (Korshin, Benjamin, and Xiao 2001).

Table 2.4
Correlations between $|\Delta A_{272}|$ and the formation of selected individual DBPs. Data for Judy Reservoir water chlorinated at varying pHs. (Adapted from Korshin et al. 2002a)

Species	pH 6.0	pH 7.0	pH 8.0
	Best-fit equation and corresponding R^2		
$CHCl_3$	$50365\|\Delta A_{272}\|^{1.97}$ $R^2 = 0.99$	$39525\|\Delta A_{272}\|^{1.83}$ $R^2 = 0.99$	$29987\|\Delta A_{272}\|^{1.70}$ $R^2 = 0.98$
DCAN	$425\|\Delta A_{272}\| - 7.5$ $R^2 = 0.97$	Decay	Decay
TCP	$315\|\Delta A_{272}\| - 2.3$ $R^2 = 0.97$	Decay	Decay
CH	$57455\|\Delta A_{272}\|^{5.46}$ $R^2 = 0.98$	$86115\|\Delta A_{272}\|^{2.59}$ $R^2 = 0.98$	$48198\|\Delta A_{272}\|^{2.43}$ $R^2 = 0.97$
MCAA	$282\|\Delta A_{272}\| - 2.4$ $R^2 = 0.92$	$291\|\Delta A_{272}\| - 3.1$ $R^2 = 0.89$	$311\|\Delta A_{272}\| - 3.6$ $R^2 = 0.90$
DCAA	$2804\|\Delta A_{272}\| - 37$ $R^2 = 0.97$	$3056\|\Delta A_{272}\| - 46$ $R^2 = 0.98$	$2859\|\Delta A_{272}\| - 40$ $R^2 = 0.99$
TCAA	$5784\|\Delta A_{272}\| - 82$ $R^2 = 0.98$	$4970\|\Delta A_{272}\| - 67$ $R^2 = 0.94$	$3501\|\Delta A_{272}\| - 45$ $R^2 = 0.98$

Subsequent research confirmed that correlations between $|\Delta A_{272}|$ and the formation of THMs, HAAs, CH, DCAN and TCP (under conditions when the latter three species were stable) were very strong and independent on the reaction time and chlorine dose. Like THMs and HAAs, these compounds were detected only when $|\Delta A_{272}|$ exceeded a threshold value; that value was different for different water qualities. A compilation of these correlations is provided in Table 2.4.

Despite the strength of the correlations shown in Table 2.4, differential spectroscopy has not been widely adopted as an approach for studying the kinetics of NOM halogenation. This might be due, in part, to the fact that the effects of bromide concentrations and temperature on the differential spectra have been studied to only a limited extent. Also, although the studies cited above indicate that differential spectroscopy can be used to probe the formation of halogenated intermediates, few attempts have been made to utilize this approach to explore the mechanisms of NOM halogenation. As the value of the information provided by differential spectroscopy becomes more widely recognized, it is anticipated that this technique will take a more prominent role in studies of NOM halogenation.

SUMMARY OF EXISTING KINETIC MODELS FOR DBP FORMATION

The publications reviewed above indicate several recent trends in the modeling of DBP formation kinetics, including consideration of kinetically distinct sites in NOM, greater attention to the formation of DBPs other than chloroform (including brominated species and HAAs), and a shift away from purely statistical models toward more mechanistic ones. Nevertheless, existing models continue to suffer from serious limitations, because key information (e.g., the identity and concentration of reactive sites and intermediates, and the nature of the rate-limiting reaction) is not known. In addition, all of the models suffer from an inadequate understanding of the stoichiometry of the reactions and the chlorine decay process.

Given the number of DBP species that are of concern and the number of water quality variables that might influence DBP formation, the task of developing a comprehensive model is daunting. However, recognition of some key features of the reactions can provide useful guidance. Specifically, a fundamental model that accounts for multiple types of reactive sites in the original NOM molecules, considers the formation of intermediates in the reaction pathways, and incorporates a rational approach to account for changes in the concentrations of reactive sites and chlorine during the reaction period is virtually the only approach that has the potential to achieve the ultimate goal. Complicated though such a process might be, in the end it is simpler and more likely to succeed than development of endless permutations of empirical equations that have little or no relationship to the underlying processes that are being modeled.

Several challenges are associated with the development of a comprehensive mechanistic model of DBP formation. The first and probably most difficult is related to the inherent complexity of NOM molecules. A second challenge derives from the large number of DBPs of interest and the effort required to carry out analyses both for the DBPs and the reactants that generate them (NOM and disinfectant). Specifically, although extensive data exist on the long-term yield of DBPs, data on DBP formation during the potentially critical period immediately after the oxidant is added are extremely scarce.

Because of the paucity of short-term DBP data and difficulties in understanding the underlying reaction pathways, models of DBP formation kinetics have been calibrated using at best a few data points related to short-term kinetics. This has led researchers to make unrealistic assumptions in the formulation of mathematical models. For instance, the vast majority of DBP formation kinetics models use only the initial NOM concentration and either the initial Cl_2 dose or the long-term Cl_2 residual as input parameters, whereas the actual values of these parameters, and especially of the rapidly reacting sites, might change dramatically at very short times after oxidant addition.

The central premise guiding the current research project was that improvements in the accuracy of DBP predictions hinge on overcoming key shortcomings in the formulation of DBP kinetics models, and that these key shortcomings are closely related to analytical limitations. Specifically, researchers have been unable to collect sufficient, accurate information about either the reactants or products of interest in the time frame of significance to develop successful models. Significant progress has been made recently in developing both a theoretical model and spectroscopic analytical tools that can partially overcome these obstacles and can provide important insights into the chemistry of DBP formation. Potentially, the model and tools could also serve a practical role in the operation of water treatment plants.

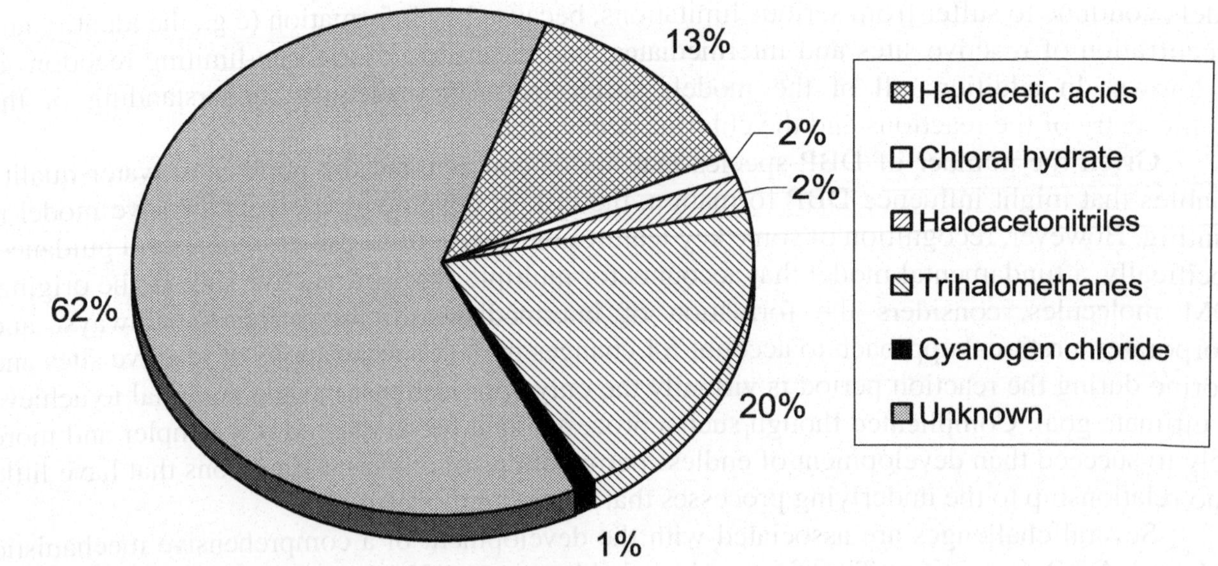

Data from Richardson, Simmons, and Rice 2002.

Figure 2.1 Typical distribution of DBP types contributing to TOX

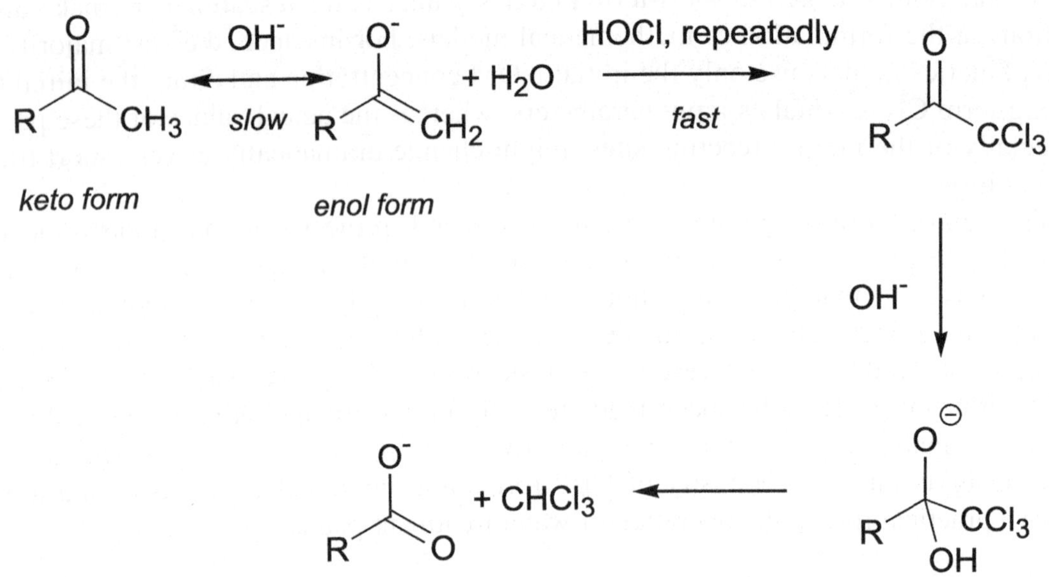

Data from Larson and Weber 1994.

Figure 2.2 Pathway of chloroform formation from a keto-enol functional group (haloform reaction)

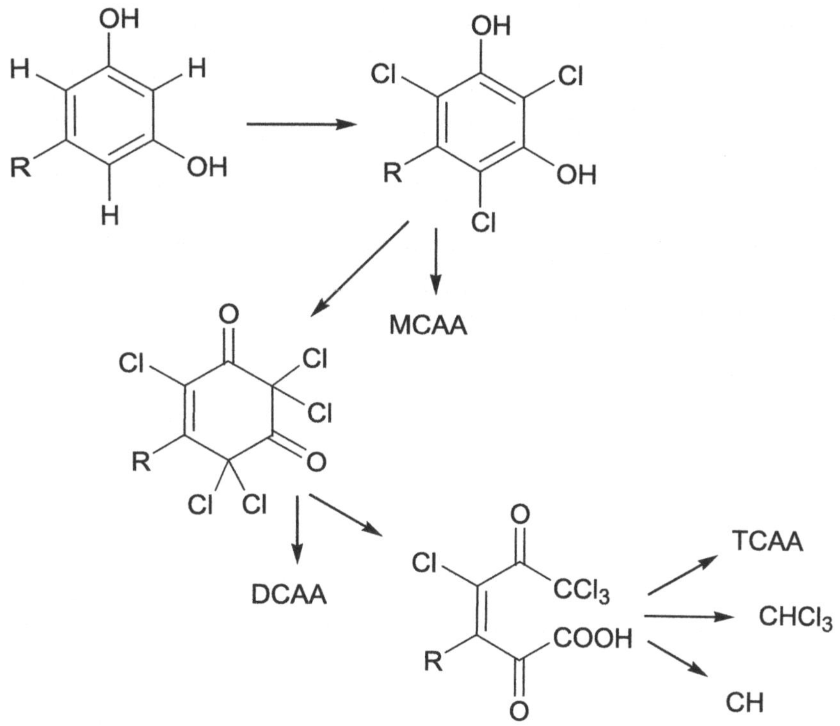

Data from Boyce and Hornig 1983.

Figure 2.3 Possible pathway for sequential attacks of free chlorine on an activated aromatic ring

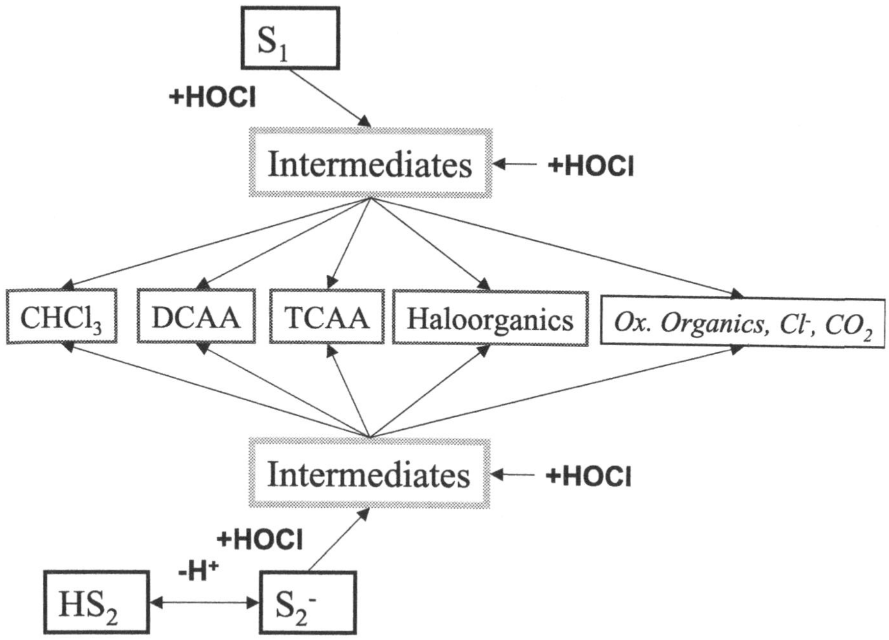

Figure 2.4 Conceptual reaction pathway for NOM chlorination, data from McClellan et al. (2000)

CHAPTER 3
MATERIALS AND METHODS

Experimental activities carried out in this project included measurements of individual DBP species (THMs, HAAs, CH, HKs, and HANs) in chlorinated "baseline" water (from Lake Washington) at variable chlorine doses, bromide concentrations, reaction times, pH and temperature. DBP formation was also analyzed after varying reaction times in chlorinated water from several other drinking water sources with a wide range of values for important water quality parameters (pH, alkalinity, bromide concentration, DOC, $SUVA_{254}$); in these latter experiments, only a single Cl_2:DOC ratio was explored. The change induced by chlorination in the absorbance spectrum of the samples (ΔA_λ vs. λ) was measured in all these experiments.

The above experiments were carried out using conventional reaction vessels with minimum reaction times of approximately 5 minutes. Additional work was carried out to quantify the kinetics of halogenation at extremely short reaction times (as short as 10 ms) using stopped-flow and quench-flow instrumentation. The data from the experiments were used to formulate and test models for the kinetics of DBP formation and for the correlations between formation of various DBPs and ΔA_λ.

MATERIALS

Laboratory Water, Reagents and Stock Solutions

Reagent water with residual DOC < 0.1 mg/L and resistivity of 18.2 MΩ-cm was obtained from a water purification system.* All chemicals were American Chemical Society (ACS) reagent grade or better. ACS-certified sodium bicarbonate ($NaHCO_3$, CAS 144-55-8), sodium tetraborate ($Na_2B_4O_7 \cdot 10H_2O$, CAS 1303-96-4) and potassium phosphate (KH_2PO_4, CAS 7778-77-0) were used to prepare buffers. Sodium perchlorate ($NaClO_4$, CAS 7601-89-0) was employed to maintain a constant ionic strength in selected experiments. Potassium bromide (KBr, CAS 7758-02-3) was used to amend samples with bromide, and a reagent-grade solution of sodium hypochlorite (NaOCl, CAS 7681-52-9) with 5.6% available chlorine was used as the source of free Cl_2.

A free chlorine stock solution was prepared by dilution of the sodium hypochlorite reagent with purified water. The solution was standardized by the DPD (N,N-diethyl-p-phenylenediamine) colorimetric method. In accordance with Standard Method 4500-Cl G (APHA, AWWA, and WPCF 1995), calibration curves were developed using a potassium permanganate solution, whose absorbance was measured at 515 nm. The standardized, stock solutions were stored at 4°C in a glass bottle wrapped in aluminum foil. In cases where it was necessary to quench the residual free Cl_2 in a solution, the quenching was accomplished with either sodium sulfite (Na_2SO_3, CAS 7757-83-7) or ammonium chloride (NH_4Cl, CAS 12125-02-9).

In some experiments, bromine (Br_2, CAS 7726-95-6) was used for halogenation. A bromine stock solution was prepared by dissolving 26 µL of liquid Br_2 (99.995% pure) in 1.0 L of ultra-pure water to form a solution of 10^{-3} M free bromine.

All solvents used in extractions were high-purity grade.

---

*Super-Q Plus, Millipore Corp., Bedford, Mass.

Containers and Glassware

Preparation of all labware for sampling of solutions followed the recommendations of Standard Methods (APHA, AWWA, and WPCF 1995). Glassware used to store water samples prior to chlorination (40-L glass bottles) was soaked in 15% nitric acid for more than one day, thoroughly rinsed with purified water, and dried. Glassware used to store and process chlorinated samples was cleaned with 4 M HNO_3 and then submerged in a solution of 5% NaOH + 2% $KMnO_4$ (by weight) to remove organic contaminants. It was then rinsed with purified water and baked in a furnace at 400°C for one hour. All vial caps were washed with a mild solution of soap and water, rinsed with purified water and HPLC-grade acetone, and then placed in a drying oven at 80°C for one hour. Volumetric glassware and pipettes that could not be heated in the oven were kept in a nitric acid bath and thoroughly rinsed with purified water before use. Disposable glass pipettes were used for chemical transfers.

Laboratory chlorination experiments were conducted in the dark, in 3.8-L poly-tetra-fluoro-ethylene (PTFE) sampling bags. These bags are chemically resistant and leak-proof and can be kept headspace-free throughout the reaction period and during sample withdrawal.

WATER SOURCES

Baseline Water From Lake Washington

The baseline water used in this research was from Lake Washington (LW), located in metropolitan Seattle, Wash. Raw water was sampled on the northwestern shore of the lake near Warren Magnuson Park. The water was filtered sequentially through two membrane filters* with 5.0- and 0.45-μm nominal pore sizes, respectively, within 24 hours of collection. Both filters were housed in polypropylene filter media cartridges. The filtered water was stored in pre-washed glass bottles and kept in the dark at 4°C.

The DOC concentration in the collected water varied somewhat during the course of the study but typically was close to 3.0 mg/L. The pH of the water was 7.5 ± 0.5. The bromide concentration in the lake water was below the 3-μg/L method detection limit (MDL) of the anionic chromatographic method employed to measure the concentrations of anions, but, as is discussed subsequently, some bromine-containing DBP species could nevertheless be quantified with confidence in the chlorinated samples.

The DOC concentration in the baseline water had to be increased for the quench-flow and stopped-flow experiments to improve the sensitivity of the tests. The pre-concentration was accomplished using a rotary evaporator† equipped with a 2.5-L round bottom flask. The process was carried out at 75–80°C, and the air pressure was lowered with a water aspirator. Following rotary evaporation, the sample was adjusted to pH 8–9, filtered through a 0.45-μm filter to remove calcium carbonate precipitate, and then acidified to pH 7. In some cases, the samples were then subjected to another rotary evaporation step to increase the DOC concentration further. The final product water, with a DOC concentration ranging from approximately 45 to 100 mg/L, was filtered through a 0.45-μm filter and stored in a PTFE container at 4°C.

---

*POLYCAP™ 36 HD and POLYCAP™ 75 AS, Whatman Inc., Clifton, N.J.

†Rotavapor R-2, Büchi Labortechnik, Flawil, Switzerland

Other Water Sources

Untreated water from sources in Florida, Iowa, Minnesota, Nebraska, South Dakota, Texas and Utah was also used in chlorination experiments. All water samples were collected by utility personnel and shipped in high-density polyethylene or polypropylene containers to the University of Washington by overnight carrier. Upon arrival, the samples were filtered though a 0.45-μm filter, after which they were stored at 4°C until use. All these samples were analyzed for dissolved inorganic and organic carbon concentration, bromide concentration, UV absorbance, and background concentrations of individual DBP species. Selected water quality parameters related to DBP formation in these samples are summarized in Table 3.1.

METHODS

Conventional Experiments Investigating Chlorination Kinetics

In the experiments investigating chlorination over time periods of several minutes or longer, chlorination was carried out in the headspace-free PTFE bags described previously. In experiments with varying bromide concentrations, the desired amount of bromide was added to the baseline water from the stock solution of potassium bromide. The pH was adjusted to 7.0 with strong acid or base and was maintained at that value with phosphate buffer. Selected experiments investigated the influence of pH on the formation of DBPs over the pH range from 6.0 to 9.0. In these experiments, mixtures of phosphate and borate buffer were used to control pH, per Standard Method 5710-B (APHA, AWWA, and WPCF 1995).

To begin the reaction, the NaOCl stock solution was injected into the bag to yield the desired Cl_2:DOC ratio. In most experiments, this ratio was 1.5 mg Cl_2 per mg DOC, although in one set of experiments, ratios from 0.1 to 3.0 mg/mg were explored. Most experiments with the baseline water lasted 7 days, and those using water from the other sources lasted 3 days. Samples were withdrawn from the PTFE bags by opening a valve in the bag and discharging liquid directly into a glass vial. Typically, several samples were taken during the first few hours, and samples were taken less frequently thereafter.

The concentration of residual free chlorine in the samples was determined immediately after they were collected. Other samples were transferred into 42-mL amber glass vials fitted with PTFE-lined caps and containing sodium sulfite at a stoichiometric ratio specified in Standard Method 5710-B to quench the residual disinfectant. The samples were then processed to determine DBP concentrations and UV absorbance. All quenched samples were stored in the dark at 4°C. All analyses were conducted within one week.

The majority of the experiments were performed at room temperature in the dark. In addition, experiments with the baseline water were carried out at temperatures of 4°C and 10°C. In these experiments, the PTFE bags with chlorinated water were placed in a constant-temperature water bath.

Quench-Flow Experiments

The quench-flow (QF) experiments were conducted to generate samples containing NOM chlorinated at reaction times between approximately 10 ms and 100 s. The apparatus used to

Table 3.1
Major water quality characteristics of source waters used in experimental work

Name (Location)	pH	Alk, mg/L as $CaCO_3$	DOC, mg/L	A_{254}, cm^{-1}	$SUVA_{254}$, $m^{-1}mg^{-1}L$	Br^-, μg/L
Lake Washington (Seattle, Wash.)	7.5	40	3.0	0.057	1.9	<3
Little Cottonwood Creek (Metropolitan Water District of Salt Lake and Sandy, Salt Lake City, Utah)	7.9	59	1.1	0.010	0.9	<3
Nueces R. (City of Corpus Christi Water Department, Corpus Christi, Texas)	8.1	150	4.2	0.098	2.4	340
Jordan R. (Jordan Valley Water Conservancy District, West Jordan, Utah)	8.1	400	4.8	0.068	1.4	233
Lake Worth (City of Ft. Worth, Ft. Worth, Texas)	7.7	110	4.9	0.090	1.7	195
Eagle Mt. Lake (Tarrant Regional Water District, Fort Worth, Texas)	8.2	128	4.7	0.088	1.9	<3
Missouri R./ Florence WTP (Metropolitan Utilities District, Omaha, Neb.)	8.2	171	4.8	0.12	2.5	<3
Missouri R./ Narrows WTP (Council Bluffs Water Works, Council Bluffs, Iowa)	8.0	180	4.7	0.12	2.6	<3
Mississippi R./St. Paul Headworks (St. Paul Water Utility, St. Paul, Minn.)	8.6	150	6.9	0.17	2.4	<3
Mississippi R. (Minneapolis Water Works, Minneapolis, Minn.)	8.3	160	10.7	0.33	3.1	<3
Big Sioux Aquifer (Sioux Falls Public Works, Sioux Fall, S.D.)	7.9	230	8.3	0.19	2.3	<3
Hillsborough R. (City of Tampa, Tampa, Fla.)*	7.2	86	27.5	1.38	4.9	92

* Hillsborough R. water was diluted 6× prior to all experiments.

generate the samples was a rapid quench-flow system*, comprising a sample handling unit (shown schematically in Figure 3.1) and an electronics support unit. The apparatus is designed to carry out the rapid mixing of reagents (in the current case, the raw water of interest and the chlorine dosing solution), and then quench the chemical reactions after a short, specified reaction period. After quenching, the samples were collected and processed using conventional techniques to analyze for the DBPs.

This routine of sample handling and preparation has both benefits and drawbacks. The main benefit, of course, is that DBP generation at very short and well-defined reaction times can be assessed. However, it must be recognized that some transformations of halogenated NOM may take place in the period between halogen quenching and analysis of the DBPs, due to the instability of some of the intermediates. The nature and extent of any such transformations remain unknown. In addition, the hydraulic design of the quench-flow apparatus is such that some mixing of the sample with the phosphate buffer solution employed to drive the sample into and out of the reaction loop is unavoidable.

All solutions employed in quench-flow experiments were loaded using 2.5-mL gas-tight, glass barrel syringes with spring-backed PTFE piston tips. All internal flow circuits were made of poly-ethyl-ethyl-ketone (PEEK) or PTFE materials. Prior to each series of experiments, the flow paths were washed and primed using the phosphate buffer solution.

The system has three syringes, which were filled with raw water, free chlorine solution, and sulfite quenching solution, as shown in Figure 3.2. Both the raw water and chlorine solutions were pre-adjusted to 7.0 using phosphate buffer. The dose of sodium sulfite used to quench the reaction was ca. 200% of the stoichiometric requirement based on the initial Cl_2 dose; the excess sulfite was required because of significant permeation of gaseous oxygen through the plastic tubing used in the internal plumbing of the instrument.

Each shot of reagents into the system produced 0.5 mL of chlorinated and quenched solution. Typically, the output from 20 shots (i.e., 10 mL) was combined before the samples were analyzed. To minimize the escape of volatile compounds during and after sample preparation, when samples were to be analyzed for DBPs, each shot of chlorinated and quenched solution was immediately injected beneath a layer of methyl tert-butyl ether (MTBE) in a closed container, in essence beginning an instantaneous extraction step; samples to be analyzed for non-volatile analytes were not subjected to this step. The composite samples were then diluted, if necessary, and analyzed for individual DBP species and UV absorbance.

Stopped-Flow Experiments

In some cases, stopped-flow (SF) experiments were carried out in parallel with the quench-flow experiments. The SF instrument employed in this project† utilizes drive rams to push pre-set amounts of the reagents into the reaction chamber, where they are rapidly mixed. The reaction chamber also serves as an optical chamber with a 1-cm path length, through which absorbance measurements can be taken at a very high frequency. A schematic of the instrument is shown in Figure 3.3.

The fundamental difference between SF and QF experiments is that, in the SF mode, the use of an optical cell as the mixing chamber and the integration of that cell into a high-precision

---

*RQF-63, Hi-Tech Scientific, Salisbury, United Kingdom

†Model SX.18MV, Applied Photophysics Ltd., SX.18MV, Leatherhead, United Kingdom

absorbance spectrophotometer allow the absorbance spectrum of the sample to be followed semi-continuously throughout the reaction period, starting at very short reaction times. On the other hand, the samples in the cell cannot easily be quenched and collected, so that this instrument is not useful for off-line DBP analysis of the chlorinated samples.

Analytical Methods

Dissolved Organic Carbon

A carbon analyzer was used for DOC analyses.* Prior to analysis, samples were filtered through a pre-washed polycarbonate 0.4-μm filter and stored in 10-mL glass test tubes that had been baked at 400°C. The instrument was calibrated using solutions of sodium biphthalate and ultra-pure water boiled with an excess of sodium persulfate as prescribed in the instrument's manual and in Standard Methods (APHA, AWWA, and WPCF 1995). The calibration was checked daily. Additionally, sodium biphthalate standards containing 5 mg/L DOC were run every ten samples. The relative standard deviation of the measurements was <10% and the reproducibility was within 5%.

Residual Chlorine

Concentrations of free chlorine in chlorinated waters were measured using the DPD Colorimetric Method (Standard Method 4500 Cl G [APHA, AWWA, and WPCF 1995]). In these measurements, the absorbance of solutions containing the DPD-active chlorine indicator was measured at 530 nm using a 2.54-cm optical cell. A UV spectrophotometer was employed in these analyses.†

Ultraviolet Absorbance

Ultraviolet absorbance was analyzed using a dual-beam spectrophotometer.‡ The absorbance was typically measured from 200 to 600 nm at 1-nm increments, using 5-cm quartz cells. The scan rate and smoothing factor were 240 nm/min and 2 nm, respectively. The instrument baseline was established using purified water and was checked every 10 measurements. In all cases, the spectra were interpreted only at wavelengths > 230 nm, where spectral interferences associated with the possible presence of nitrate, sulfite and bromide are negligible.

Trihalomethanes and Other Volatile DBPs

THMs and other volatile DBPs were analyzed by liquid-liquid extraction according to EPA Method 551.1 (USEPA 1995a) using a gas chromatograph§ (GC) equipped with an electron capture detector (ECD) and a fused silica capillary column (30 m × 0.53 mm i.d., 1.5 μm film thickness).** A 30-mL sample aliquot was extracted with 3 mL of MTBE, and 1.0 μL of the

---

*Model 700 or, later in the project, Model 1010, OI Analytical, College Station, Texas

†Model DR/4000, HACH Corporation, Loveland, Colo.

‡Model Lambda-18 UV/VIS Perkin-Elmer Gmbh., Überlingen Germany

§Perkin-Elmer AutoSystem XL, Perkin-Elmer Corporation, Boston, Mass.

**Model DB-5, Agilent Technologies, Palo Alto, Calif.

extract was then injected into the GC. The temperature program used to separate the analytes was as follows: the initial temperature of 35°C was held for 12 min, after which the temperature was increased at a rate of 8°C/min to 190°C, where it was held for 3 min. The temperatures of the injector and detector were 200°C and 350°C, respectively.

The instrument was calibrated using conventional calibration standards. Five aqueous calibration standards bracketing the analyte concentration range expected in the samples were analyzed together with each set of samples. All samples were analyzed in duplicate within one week of extraction. Internal standards were used with some samples to check for the extraction efficiency. The relative standard deviation of the internal standards was approximately 20%.

Haloacetic Acids

All nine chlorinated and/or brominated HAAs were analyzed using EPA Method 552.2 (USEPA 1995b). This procedure involves liquid-liquid extraction and subsequent sample derivatization and determination of volatile products by gas chromatography with ECD detection. A 30-mL sample was adjusted to pH < 0.5 with 50% sulfuric acid and extracted with 4 mL of MTBE. The HAAs that were extracted into the MTBE phase were converted to their methyl esters by the addition of acidic methanol followed by slight heating. The acidic extract was neutralized by a back-extraction with a saturated solution of sodium bicarbonate. The target analytes were identified and analyzed using the same GC as was used for THM analysis (described previously). The temperature program was as follows: the initial temperature of 35°C was held for 12 min, after which the temperature was increased at a rate of 5°C/min to 200°C, where it was held for 2 min.

The method detection limits (MDLs) and other relevant data for all the DBPs analyzed in the project are shown in Table 3.2. The MDLs were calculated based on the analysis of seven fortified samples, according to the following equation:

$$\text{MDL} = t_{n-1,\,0.99}\,\sigma$$

where $t_{n-1,\,0.99}$ is the students t value appropriate for a 99% confidence level and a standard deviation estimate with n–1 (i.e., six) degrees of freedom, and σ is the standard deviation of the replicate analyses.

Although the MDLs shown in Table 3.2 were experimentally obtained, the DBP analyses were carried out by several different researchers during the course of the project, and some inconsistency among researchers is inevitable. Therefore, for the purposes of this report, effective MDLs are adopted that are triple the values shown in the table.

Total Organic Halide (TOX)

TOX measurements were performed with a TOX analyzer* using the absorption-pyrolysis-titrimetric method (Standard Method 5320-B [APHA, AWWA, and WPCF 1995]). One or two drops of concentrated nitric acid was added to each 25- or 50-mL sample to decrease the pH to ~1.5. The acidified sample was passed through two glass columns in series, each packed with

---

*Model TOX-10Σ, Mitsubishi Chemical Corporation, Tokyo, Japan

Table 3.2
Target compounds analyzed, corresponding abbreviations, boiling points and method detection limits (in µg/L)

Analyte	Abbreviation	BP (°C)	MDL*	MW
	Method 551.1			
Chloroform	$CHCl_3$	62	0.08	119.37
Trichloroacetonitrile	TCAN	86	0.01	144.39
Dichlorobromomethane	$CHBrCl_2$	90	0.19	163.82
Chloral hydrate	CH	98	0.02	165.38
Dichloroacetonitrile	DCAN	113	0.09	109.93
1,1-Dichloro-2-propanone	DCP	120	0.03	126.94
Chloropicrin	CPN	112	0.01	164.37
Dibromochloromethane	$CHBr_2Cl$	120	0.12	208.27
Bromochloroacetonitrile	BCAN	127	0.01	154.39
1,1,1-Trichloro-2-propanone	TCP	149	0.02	161.39
Bromoform	$CHBr_3$	150	0.18	252.72
Dibromoacetonitrile	DBAN	168	0.04	198.84
	Method 552.2			
Monochloroacetic acid	MCAA	189	0.27	94.47
Monobromoacetic acid	MBAA	208	0.13	138.92
Dichloroacetic acid	DCAA	194	0.16	128.93
Trichloroacetic acid	TCAA	197	0.10	163.38
Bromochloroacetic acid	BCAA	~211	0.11	173.38
Dibromoacetic acid	DBAA	233	0.07	217.83
Bromodichloroacetic acid	BDCAA	n.a.†	0.18	207.83
Dibromochloroacetic acid	DBCAA	n.a.	0.23	252.28
Tribromoacetic acid	TBAA	245	0.22	296.73

*Method detection limit calculated based on seven replicate laboratory-fortified blanks; as noted in the body of the report, "effective MDLs" equal to triple the values shown are adopted throughout the report
†n.a.: not available

activated carbon to adsorb halogenated organic species. Inorganic halides retained by the activated carbon were displaced by washing the column with potassium nitrate solution. The activated carbon was pyrolyzed in a quartz furnace in a stream of oxygen and carbon dioxide. The halogenated organics were thereby transformed to corresponding volatile inorganic halides and carried by the gas stream into an electrochemical titration cell. The cell used a silver working electrode and a silver-sensitive reference electrode to quantify the halides by coulometric titration. The recovery of TOX was 100±5% based on tests using 2,4,5-trichlorophenol, and the sensitivity of the instrument was 10±5 ng.

Strictly speaking, concentrations of organic halides evaluated based on pyrolysis of activated carbon that retains them should be referred to as adsorbable organic halides (AOX) rather than TOX. However, the difference between total organic halides and AOX is believed to be minor in drinking water; therefore, for the sake of consistency with many other publications dealing with DBP formation in potable water, the latter term (TOX) is employed in the remainder of this report.

Inorganic Anions

An ion chromatograph* equipped with a conductivity detector, gradient pump, and separation column, was used in the analyses of all inorganic anions. The flow rate was 0.25 mL/min, and a 5% (191 mM) NaOH solution in purified water was used as eluent.

*Model DX-500 chromatograph, with CD20 conductivity detector, GP40 gradient pump, and AS-11 separation column; all from Dionex Corp., Sunnyvale, Calif.

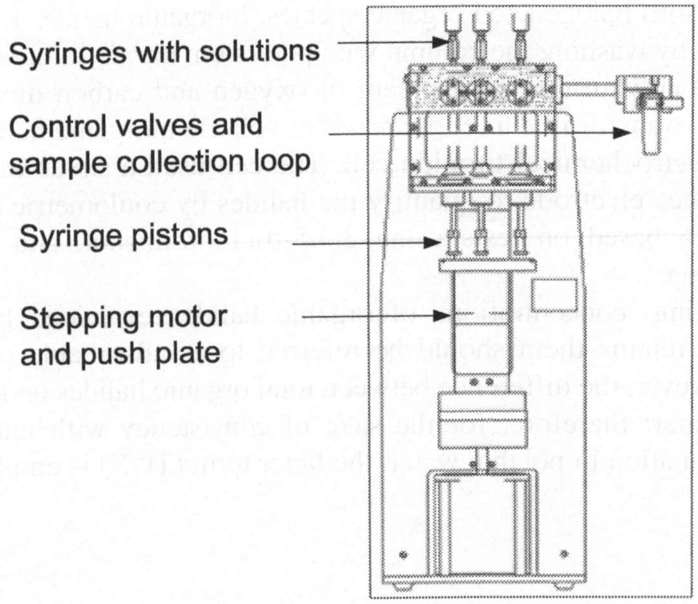

Source: Hi-Tech Ltd. (2003).

Figure 3.1 Main components of the sample preparation unit of the RQF-63 quench-flow system

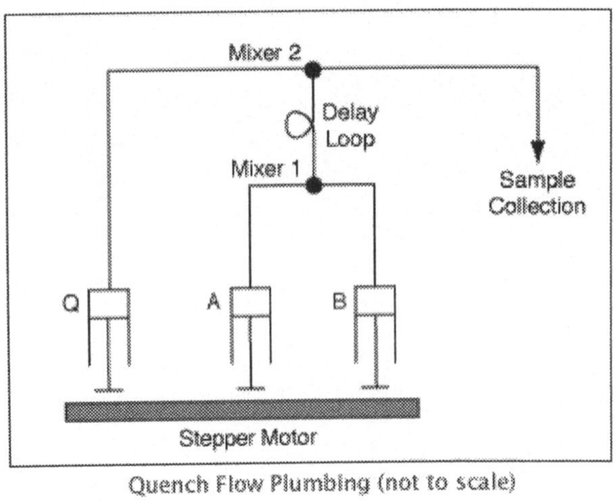

Source: Hi-Tech Ltd. (2003).

Figure 3.2 Schematic of the internal plumbing in the RQF-63 quench-flow system. Syringes Q, A and B hold solutions of the quenching agent (sodium sulfite), NOM and chlorine, respectively.

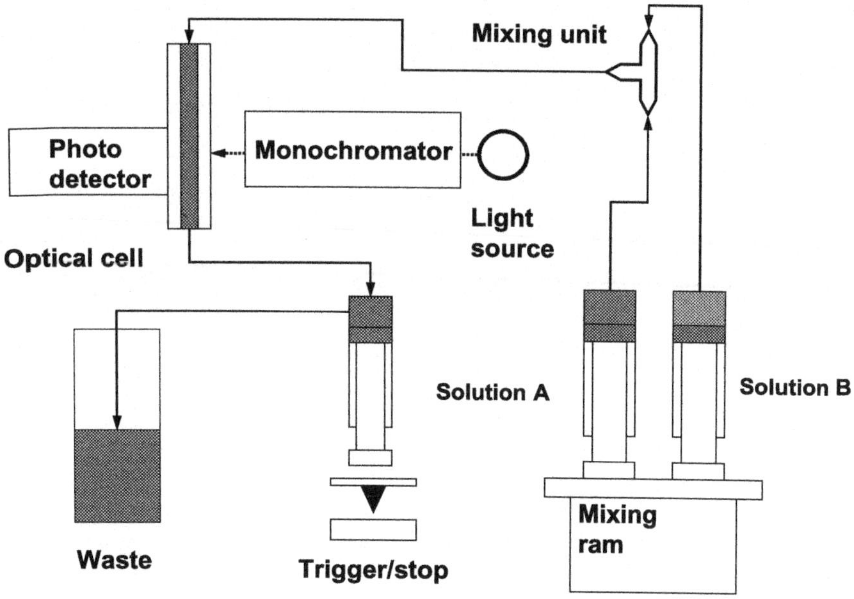

Figure 3.3 Schematic representation of a stopped-flow experiment

CHAPTER 4
KINETICS OF DBP FORMATION IN THE BASELINE WATER

INTRODUCTION

In this chapter, the results of conventional kinetic experiments, with reaction times from 5 min to 7 d, are presented for chlorination of baseline (LW) water. The data include DBP formation as a function of reaction time for systems dosed with varying amounts of free Cl_2 (corresponding to varying Cl_2:DOC ratios), and also varying with respect to pH, temperature, and Br^- concentration. This chapter also includes data collected in stopped-flow and quench-flow experiments, demonstrating the reaction kinetics at very short reaction times (from < 1 s to 100 s). Data obtained in the experiments with water from other sources are presented in Chapter 5, in conjunction with the modeling efforts.

The results presented in this chapter comprise the core data set for developing and testing kinetic models in Chapter 5. The data are presented here with very limited interpretation, since the primary objective of the study was to improve the modeling of DBP formation kinetics, not to explore DBP formation chemistry. Because one of the hypotheses being tested is that careful monitoring of $|\Delta A_{272}|$ over time can greatly reduce the need for detailed kinetic models of DBP formation, correlations between $|\Delta A_{272}|$ and DBP formation in the systems of interest are also presented; interpretation of these relationships is presented in Chapter 6.

UV ABSORBANCE, CHLORINE CONSUMPTION, AND DBP FORMATION KINETICS IN BASELINE CHLORINATION EXPERIMENTS

Figure 4.1 shows the UV absorbance of chlorinated LW water as a function of reaction time. The uppermost line is for the raw water (i.e., before chlorination). Absorbance decreased over time, especially between 260 and 300 nm. The corresponding differential absorbance spectra ($|\Delta A_{272}|$ vs. λ) are shown in Figure 4.2.

Figure 4.3 shows the time profile of $|\Delta A_{272}|$ in these experiments. After only 5 min of reaction (the first data point obtained), $|\Delta A_{272}|$ was already substantial (0.0079 cm^{-1}, or approximately 28% of the value at $t = 7$ d), and within 1 h, the value of $|\Delta A_{272}|$ had reached approximately 39% of its 7-d value.

The residual chlorine concentration in the same experiment is shown in Figure 4.4. The time profile of Cl_2 consumption was similar to that of $|\Delta A_{272}|$; i.e., Cl_2 consumption was significant by the time the first data point was collected, and it increased continuously, but at a steadily declining rate, thereafter. As a result, $|\Delta A_{272}|$ and Cl_2 consumption were well correlated with one another ($R^2 = 0.96$), as shown in Figure 4.5. The fact that the extrapolation of the line intersects the x axis at a finite value of $|\Delta A_{272}|$ indicates that the initial reactions of the NOM with chlorine destroy more UV absorbance (per mg of Cl_2 consumed) than do the reactions that occur later.

Figure 4.6 characterizes the generation of THM and HAA species in this experiment. Although the range of concentrations of these DBPs is very large, the kinetics of formation of the individual compounds are quite similar to each other and to the kinetic patterns described above for $|\Delta A_{272}|$ development and Cl_2 consumption. Presumably, this pattern reflects the fact that, when chlorination starts, the most reactive sites react rapidly, and a continuum of less reactive sites reacts thereafter.

The concentration of $CHCl_3$ formed in these experiments is shown as a function of $|\Delta A_{272}|$ in Figure 4.7. The data can be fit by a straight line ($R^2 = 0.97$) that does not pass through the origin, consistent with results for similar systems studied previously (Korshin, Li, and Benjamin 1999; Korshin et al. 2002a,b). This behavior has been interpreted as support for the hypothesis that chloroform is the end product of a series of reactions, one or more of which destroy UV absorbance before subsequent ones generate the DBP. Note that, since the data must pass through the origin (a value of zero for $|\Delta A_{272}|$ corresponds to no reaction, and hence no chloroform formation), the relationship must be curvilinear between the origin and the data shown. The concentration vs. $|\Delta A_{272}|$ correlations for other DBPs are shown in subsequent portions of this chapter.

Effect of Cl_2-to-DOC Ratio

Development of $|\Delta A_{272}|$ and Cl_2 Consumption

The trends in $|\Delta A_{272}|$ and chlorine consumption in experiments with LW water dosed with six different Cl_2 concentrations (corresponding to Cl_2:DOC ratios from 0.1 to 3.0 mg Cl_2/mg DOC) are shown in Figures 4.8 through 4.10. In all cases, the trends are similar to those described above for the system with a Cl_2:DOC ratio of 1.5. Despite the wide range of experimental conditions, the relationship between Cl_2 consumption and $|\Delta A_{272}|$ from all the experiments can be fit quite well by a single, straight line (Figure 4.10).

DBP Formation and Its Relationship to $|\Delta A_{272}|$

Figure 4.11 shows the kinetics of formation of six DBPs (two THMs, three HAAs, and CH) in the experiments with the different Cl_2:DOC ratios. The corresponding relationships between DBP formation and $|\Delta A_{272}|$ are shown in Figure 4.12. When plotted in this latter way, most of the data for the six different Cl_2:DOC ratios collapse to a fairly narrow, uniform curve for each DBP (the data for MCAA are more scattered than for the other DBPs, and even for the other DBPs, a few obvious outliers are present).

Even though the data for several of the DBPs can be fit empirically by straight lines with a positive intercept on the $|\Delta A_{272}|$ axis, logic demands that the data pass through the origin, as noted previously. Therefore, the DBP vs. $|\Delta A_{272}|$ relationships are most accurately described as curvilinear and concave upward, at least at low to moderate values of $|\Delta A_{272}|$. Such a trend is very clear in the data for chloral hydrate (CH).

Effect of pH

Plots showing $|\Delta A_{272}|$ and Cl_2 consumption in systems in which the Cl_2:DOC ratio was fixed at 1.5 and the pH was varied from 6 to 9 (buffered with either phosphate or borate) are shown in Figures 4.13 through 4.15. The $|\Delta A_{272}|$ vs. time profiles in this series of experiments were almost identical at pH 6, 7, and 8, but at pH 9 the data were more scattered and the magnitude of $|\Delta A_{272}|$ was generally less than in the more acidic solutions (Figure 4.13). On the other hand, the profiles of Cl_2 consumption vs. time were nearly identical at pH 7, 8, and 9, while slightly less Cl_2 was consumed at pH 6 (Figure 4.14). The detailed mechanisms leading to these trends cannot be established from these data, but the underlying cause is undoubtedly related to the changing speciation of the chlorine species (pK_a for HOCl is 7.5), and perhaps the changing speciation of the NOM as well. The combination of these trends leads to roughly parallel, linear, but not identical plots of Cl_2 consumption vs. $|\Delta A_{272}|$ at the different pH's (Figure 4.15).

Consistent with the results obtained by many previous researchers studying other waters, $CHCl_3$ formation increased and TCAA concentration decreased with increasing pH (Figure 4.16). By contrast, in these experiments, MCAA and DCAA formation were affected minimally, if at all, by pH. CH behaved like MCAA and DCAA in this respect; i.e., no trend with pH could be discerned in the CH data. Small amounts of $CHCl_2Br$ were detected at all pH's studied, and the concentration vs. time profiles for this DBP were essentially identical at pH values of 7, 8, and 9; less of this DBP was formed at pH 6.

Figure 4.17 shows the relationships between DBP formation and $|\Delta A_{272}|$ in these experiments. For the THMs and CH, the data set migrates toward higher concentrations with increasing pH at a given value of $|\Delta A_{272}|$, whereas no significant trend is observed for the HAAs.

Effect of Bromide on DBP Speciation and Formation Kinetics

Experiments in which up to 2.0 mg/L bromide ion (Br^-) was added to the raw water prior to chlorination were conducted at a pH of 7.0, a Cl_2:DOC ratio of 1.5, and 25°C. The development of $|\Delta A_{272}|$ over time in the experiments with ≤0.2 mg/L Br^- was essentially identical to that in the control system with no Br^- added (Figure 4.18); at higher Br^- concentrations, the magnitude of $|\Delta A_{272}|$ was slightly greater.

Increasing the Br^- concentration leads to greater consumption of free chlorine to oxidize that Br^-. However, once the Br^- is converted to HOBr, it might substitute into NOM molecules or oxidize portions of those molecules, thereby reducing the number of available precursor sites that could otherwise react with free chlorine. As a result, increasing Br^- concentration is expected to increase Cl_2 consumption, but at a ratio of less than one mole of additional Cl_2 consumption per mole of Br^- in the original solution.

Figure 4.19a shows the chlorine consumption profiles in these experiments. The total amount of chlorine consumed increased with increasing Br^- concentration. While the qualitative trend is fairly clear, the data have too much scatter, and the increment in Cl_2 consumption is too small, to provide a strong quantitative correlation between the increase in Cl_2 consumption and the concentration of bromide added. The differences among the systems with different amounts of Br^- added appeared almost instantly (Figure 4.19b), suggesting that they were primarily associated with the initial oxidation of Br^- to HOBr.

A composite plot showing chlorine consumption as a function of $|\Delta A_{272}|$ in all these experiments is shown in Figure 4.20. The trend is nearly independent of Br^- concentration.

DBP formation in the experiments with varying Br^- additions is shown in Figure 4.21. As would be expected, for the THMs, increasing Br^- concentration favored formation of progressively more highly brominated DBPs. $CHCl_2Br$ and $CHClBr_2$ concentrations both increased with increasing Br^- concentration to a point, and then decreased with further increases in Br^- concentration, as the more highly brominated $CHBr_3$ species formed preferentially. A similar pattern held true for the HAAs.

Figure 4.22 shows the relationships between the concentrations of various DBPs and $|\Delta A_{272}|$ in the systems with Br^- added. As in the absence of added Br^-, the relationship are curvilinear and concave upward, and the data for MCAA are more scattered than for the other DBPs. Addition of Br^- displaces the curves upward or downward, depending on the amount of bromine incorporated into the DBP. The total, molar concentrations of THMs at a given value of

$|\Delta A_{272}|$ increased slightly with increasing Br^- concentration (Figure 4.23a); no such trend is apparent for THAAs (Figure 4.23b).

Figure 4.24a shows the amount of organic Br incorporated into identifiable brominated DBPs (THMs, HAAs, and HANs) over time for all six Br^- concentrations tested, and Figure 4.24b shows the distribution of organic Br in these species after 72 hr of reaction. These plots suggest that bromine incorporation into the DBPs followed a pattern similar to that for chlorine incorporation, and that the distribution of organically-bound bromine among the various DBPs was approximately constant over time. The fraction of the total bromine that could be accounted for by these three groups of DBPs decreased steadily with increasing Br^- concentration. For instance, after 72 hr of reaction, 50% of the bromine was present in these three groups in the system with 0.1 mg/L Br^- added, whereas only 17% was in the three groups in the system with 2 mg/L added.

Effect of Temperature

The effect of temperature on the halogenation reactions was explored in experiments at 4°C, 10°C, and 25°C in solutions with 0, 0.2, or 2 mg/L Br^- added. The pH of these systems was 7.0, and the Cl_2:DOC ratio was 1.5.

The patterns of $|\Delta A_{272}|$ development and chlorine consumption in the experiments at 4°C and 10°C with no Br^- added were similar to those in the experiments at 25°C described previously, but the rate of change of both parameters declined with decreasing temperature (Figure 4.25). Similar results were obtained with solutions with 0.2 and 2.0 mg/L Br^- added (data not shown).

The effect of temperature on formation of individual DBPs is shown in Figure 4.26. As might be expected, lowering the temperature led to a decrease in the concentrations of $CHCl_3$ and CH that formed. However, contrary to expectations, the HAA concentrations were often larger at 4°C or 10°C than at 25°C. Additional experiments are required to confirm this result and, if it is confirmed, to explore its causes and significance.

Figures 4.27 and 4.28 display the relationships of TTHM and THAA (expressed on a molar basis) with $|\Delta A_{272}|$ as a function of temperature at various Br^- doses. Temperature did not affect these relationships significantly, regardless of the Br^- concentration, except that, at a given $|\Delta A_{272}|$, THAA was noticeably less at 25°C than at the lower temperatures. The relationships between the concentrations of individual DBPs formed and $|\Delta A_{272}|$ at the three different temperatures were similar to those for the corresponding composite parameters (data not shown). The relationship between total identifiable organic bromine and $|\Delta A_{272}|$ was also relatively insensitive to temperature (Figure 4.29).

Summary of Key Results From Conventional Kinetics Experiments

The experiments described above provide a consistent data base for kinetic modeling of DBP formation following chlorination of a typical fresh water. The results are, in general, consistent with those from many previous studies; their value is thus not in their novelty, but rather in the range of water quality conditions studied in a consistent manner. The key results of these conventional kinetic experiments can be summarized as follows.

1. Chlorine consumption, DBP formation, and the decrease in absorbance at 272 nm (corresponding to an increase in $|\Delta A_{272}|$) all follow qualitative patterns consisting of a very rapid initial change, followed by a more gradual change that continues for at least several days.
2. The decrease in absorbance at 272 nm (increase in $|\Delta A_{272}|$) can be used as a master variable to monitor the progress of Cl_2 consumption and DBP formation.
3. Chlorine consumption and DBP formation are both approximately linearly related to $|\Delta A_{272}|$ after the reaction has proceeded beyond an initial stage. In this first stage of the reaction, more UV absorbance is destroyed per unit amount of Cl_2 consumed or per unit amount of DBP formed than is the case later in the process. As a result, when the linear portions of plots showing Cl_2 consumed or DBPs generated as a function of $|\Delta A_{272}|$ are extrapolated, the extrapolation intersects the $|\Delta A_{272}|$ axis at a positive value.
4. In LW water, as in many other waters studied previously, formation of $CHCl_3$ increases, and that of TCAA decreases, as the pH is increased from 6 to 9. The formation of other DBPs (DCAA, MCAA, CH) in LW water is insensitive to pH.
5. For otherwise identical conditions, increasing the Br^- concentration leads to a slight increase in chlorine consumption, and a shift to favor formation of more highly brominated DBPs.
6. For chlorination of LW water, changing the temperature over the range from 4°C to 25°C has relatively minor effects on reaction kinetics and DBP formation. THM formation in this water is slightly more sensitive than HAA formation to temperature.

SHORT-TERM KINETICS STUDIES USING STOPPED-FLOW AND QUENCH-FLOW INSTRUMENTATION

Oxidant Consumption and Changes in Absorbance

Stopped-flow and quench-flow experiments were conducted with the baseline (LW) water to assess the reaction of NOM with halogens at very short reaction times and to compare it to that observed in conventional kinetic experiments. As noted in Chapter 3, the stopped-flow and quench-flow experiments were carried out using pre-concentrated LW NOM.

In the stopped-flow experiments, the initial solution in the reaction chamber contained 22.5 mg/L DOC and 10 mg/L Cl_2 (after accounting for the dilution of each reagent by the other). The decline in absorbance associated with both the NOM and the chlorine indicated that the reaction proceeded very rapidly, with almost 70% of the chlorine being consumed within 100 s (Figure 4.30). When the same experiment was carried out using Br_2 as the oxidant instead of Cl_2 (at the same molar concentration), approximately 30% of the Br_2 was consumed before the first data point was collected (after 0.1 s of reaction). This was substantially more oxidant consumption than when the solution was dosed with Cl_2. Thereafter, the rates of oxidant consumption were similar in the two solutions. When plotted on a semi-logarithmic plot, the consumption kinetics for each oxidant could be represented as a straight line at times less than approximately 1 s, and as a somewhat steeper straight line at longer reaction times (Figure 4.31).

The differential spectra for these experiments are shown in Figures 4.32 and 4.33. Although the precision of absorbance measurements with the stopped-flow instrument was lower than that for a conventional dual-beam spectrophotometer, the shape of all the spectra in these figures is unmistakably similar to those generated in the conventional mode. In particular, the sign of the differential absorbance is negative at all wavelengths and times, the maximum of $|\Delta A_\lambda|$ is located near 272 nm, and the magnitude of $|\Delta A_\lambda|$ increases monotonically with time. The main difference between the differential spectra for chlorinated and brominated LW NOM is that, in the former case, $|\Delta A_{272}|$ increased gradually, while in the latter it was quite large at the shortest reaction time (0.1 s) and increased only slightly thereafter.

The kinetic profiles of $|\Delta A_{272}|$ development for chlorinated and brominated LW water, shown in Figure 4.34, are qualitatively similar to those for oxidant consumption. At values up to approximately 4.5 mg/L Cl_2 consumed and $|\Delta A_{272}|$ equal to 0.15 cm^{-1}, the relationship between $|\Delta A_{272}|$ and chlorine consumption (Figure 4.35) is similar to that obtained from the conventional experiments, in that it is approximately linear and, when extrapolated, passes through the origin. However, at higher parameter values, the slope of the plot decreases noticeably. By contrast, the plot of $|\Delta A_{272}|$ vs. bromine consumption is linear throughout the parameter space explored, with a slope comparable to that in the systems with Cl_2 at higher parameter values, and the extrapolated line does not pass through the origin.

The different patterns observed in the chlorinated and brominated systems are indicative of a more rapid breakdown of the chromophores when NOM is brominated than when it is chlorinated. This result is consistent with data for DBP formation, both in the current study and others reported in the literature (e.g., Cowman and Singer 1996; Nokes, Fenton, and Randall 1999).

DBP Formation

The stopped-flow experiments allowed the kinetics of halogen consumption and the differential absorbance to be probed at short reaction times, but they could not be used to assess DBP formation kinetics. Such information was obtained from quench-flow experiments. Like the stopped-flow experiments, the quench-flow experiments used pre-concentrated LW NOM. In these experiments, the initial DOC and chlorine concentrations in the solution (i.e., the concentrations immediately after the NOM-containing solution and the oxidant were mixed), were 25 mg/L and 37.5 mg/L as Cl_2, respectively.

The differential spectra of the chlorinated LW NOM at reaction times ranging from 0.05 to 60 s are shown in Figure 4.36. Although the intensity of these spectra is less than that for comparable systems at longer reaction times, the main features of the spectra (e.g., their shape, and the sign and position of the maximum) are virtually indistinguishable from those in the conventional experiments.

The kinetics of development of $|\Delta A_{272}|$ and of the formation of individual DBPs in these experiments are shown in Figure 4.37 and Figure 4.38 through 4.41, respectively. The differential absorbance developed rapidly, growing to a significant value at $t = 10$ s and remaining approximately constant from that time up to 1 min. By contrast, although all of the DBPs could be detected after even very short reaction times, the concentrations of most of the DBPs (with the sole exception of 1,1-dichloropropanone [DCP]) increased steadily throughout the first minute of reaction.

These observations support the idea that fast chlorine incorporation reactions occur prior to the release of individual DBPs. As noted previously, this idea is logical and has been proposed by other researchers, supported by various types of experimental data (Adin et al. 1991; Nokes, Fenton, and Randall 1999; McClellan et al. 2000). What is noteworthy about the present findings is that they demonstrate that the patterns that characterize long-term DBP release and speciation can be observed even at reaction times less than 1.0 s. This observation suggests that, even though NOM molecules contain sites with dramatically different kinetic properties with respect to chlorine attack, those sites are fundamentally similar in terms of the reaction pathways along which they proceed.

This conclusion is supported by the correlations between the concentrations of DBPs formed and the differential absorbance in the quench-flow experiments (Figure 4.40 through 4.43). For all the DBPs analyzed, these correlations were similar to those obtained in conventional experiments. The only significant difference was that the threshold value of $|\Delta A_{272}|$ below which DBP species were not detected (~0.006 cm^{-1}) was larger in the quench-flow experiments, a result that can be attributed to the higher DOC concentration in those experiments. The threshold $|\Delta A_{272}|$ value was approximately the same for all the DBPs analyzed, supporting the hypothesis that the DBPs all originated from the same site (or sites) in the NOM molecules.

Conclusions From Stopped Flow and Quench-Flow Experiments

The results of the stopped-flow and quench-flow experiments support the following conclusions. First, the kinetics of halogen incorporation into NOM and loss of absorbance due to breakdown of the attack sites are very rapid, so that examining the reactions even at times <1 s does not necessarily simplify the analysis significantly. Since halogen consumption and corresponding changes in absorbance and DBP formation are known to proceed over time frames extending to at least several days, NOM must contain halogen attack sites with characteristic reaction rates differing by several orders of magnitude.

Second, the release of individual DBP species is always delayed compared to the onset of halogen incorporation and absorbance destruction, thus confirming the multi-step nature of the DBP formation reactions.

Finally, the differential absorbance spectra generated by chlorination of NOM and the speciation of the DBPs formed are remarkably similar for sites that have dramatically different kinetic behavior, suggesting that the reaction pathways for these different sites are similar. Stated differently, the reaction sites in NOM can be envisioned to have reactive moieties that are very similar structurally (most likely, activated aromatic groups or unsaturated keto groups) whose reactivity varies widely. The variation in reactivity might be caused by differences in topochemistry (e.g., positions of the attack sites outside or inside the hydrophobic core of the organic quasi-polymeric molecules) or the identity of attached functional groups. All of these sites tend to release DBPs according to a consistent pattern with respect to the distribution of DBP species and the dependence on $|\Delta A_{272}|$.

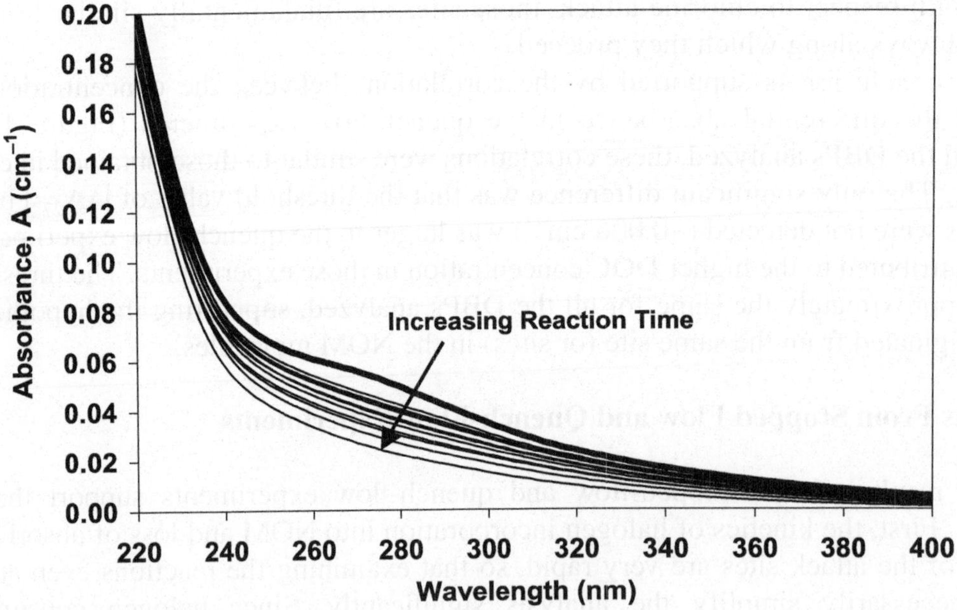

Figure 4.1 UV absorbance spectra of LW water at reaction times ranging from 5 min to 7 d, after chlorination at a Cl_2:DOC ratio of 1.5 mg/mg. pH 7, 25°C.

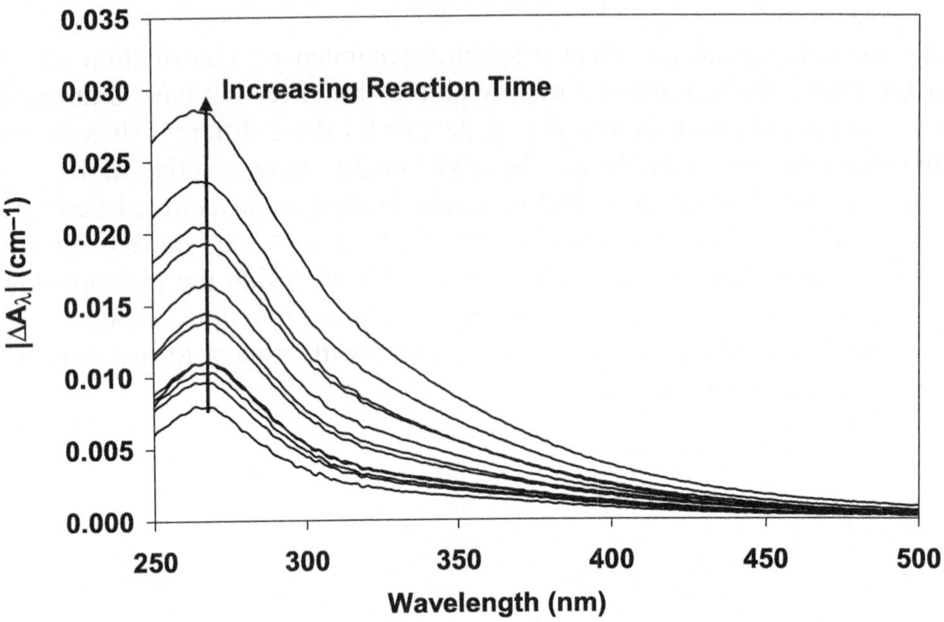

Figure 4.2 Differential absorbance spectra of the solutions shown in Figure 4.1

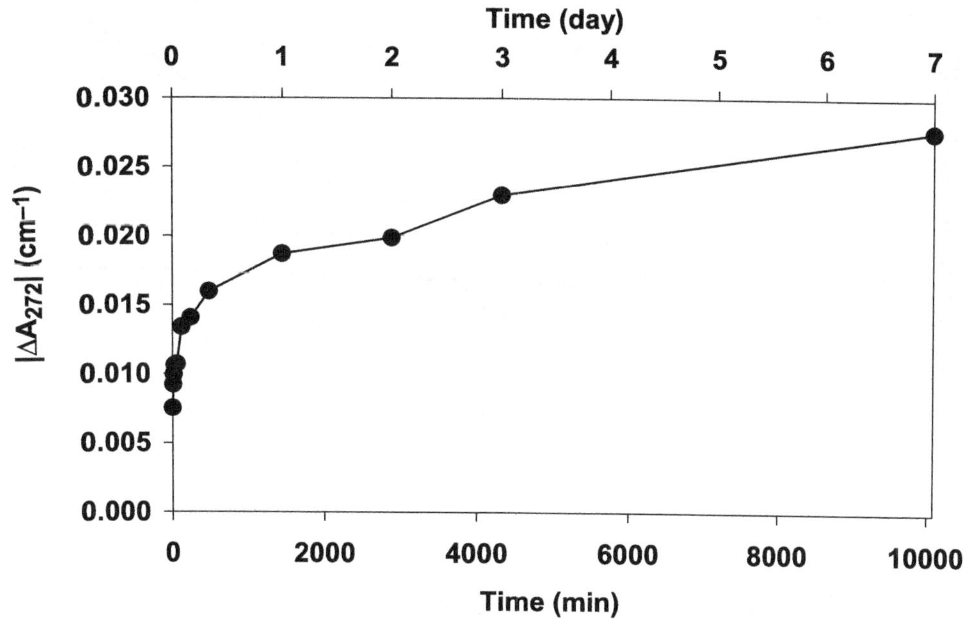

Figure 4.3 Development of the differential absorbance at 272 nm during chlorination of LW water. Conditions same as in Figure 4.1.

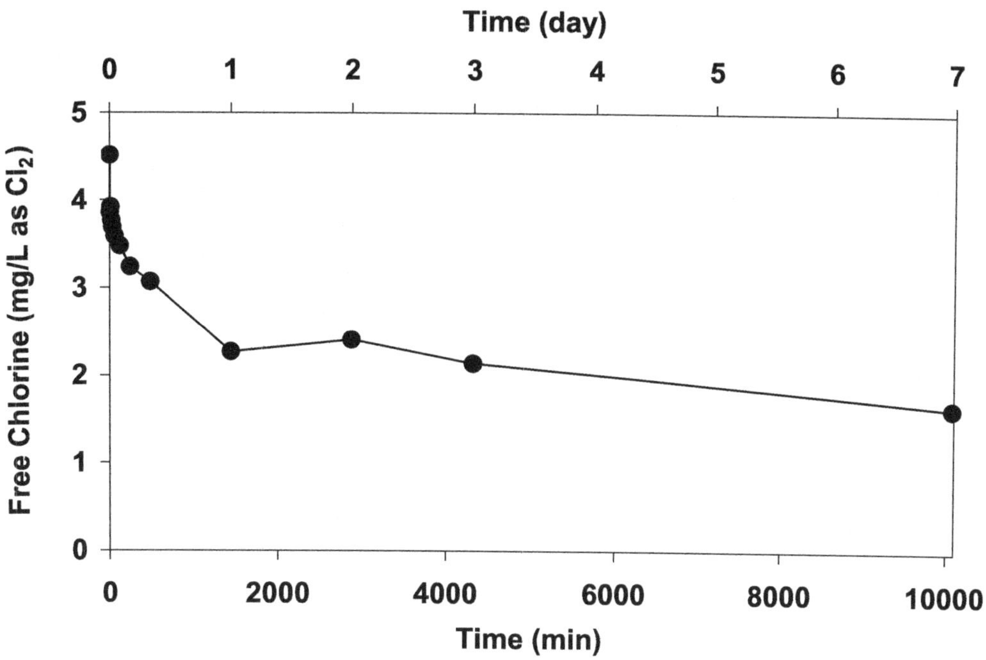

Figure 4.4 Depletion of Cl_2 during the chlorination of LW water. Conditions same as in Figure 4.1.

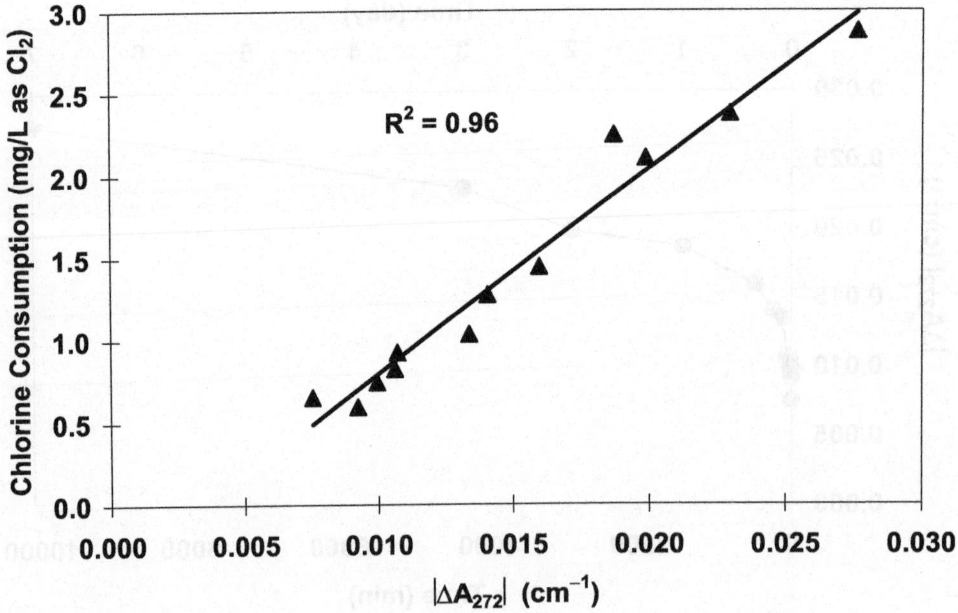

Figure 4.5 Correlation of $|\Delta A_{272}|$ with chlorine consumption in baseline water. Conditions same as in Figure 4.1.

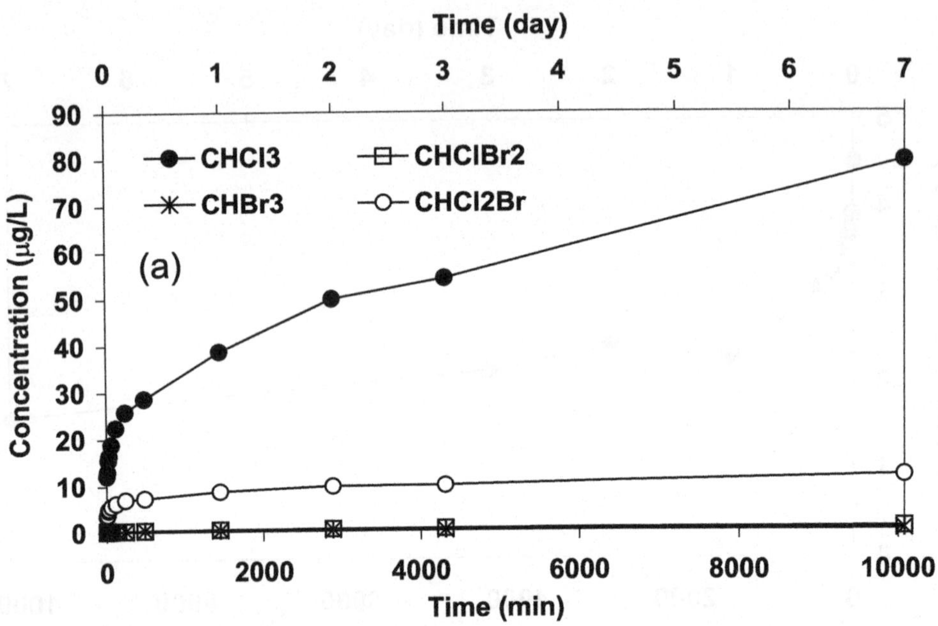

Figure 4.6a THM and HAA formation in LW water at reaction times from 5 min to 7 d. Conditions same as in Figure 4.1. (a) THMs.

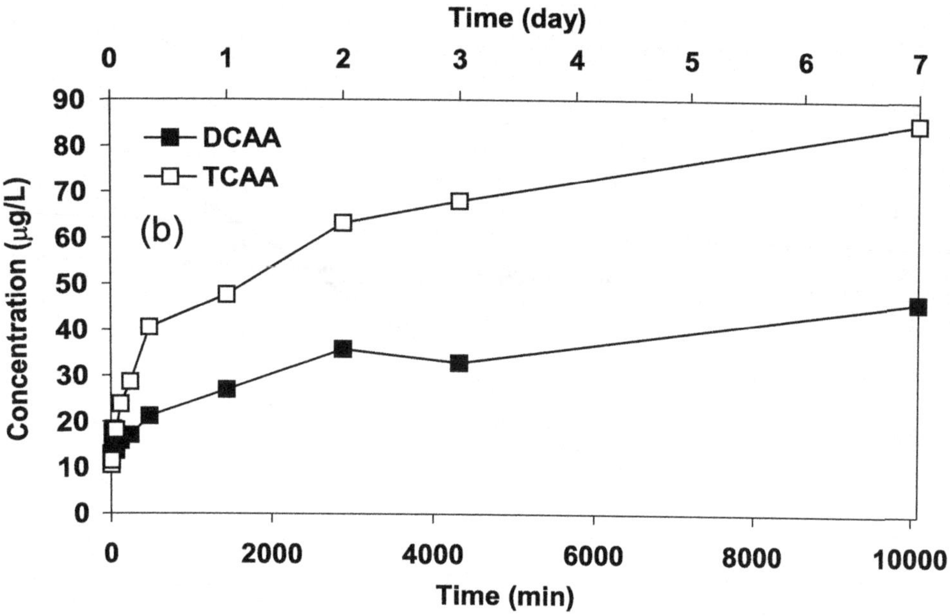

Figure 4.6b THM and HAA formation in LW water at reaction times from 5 min to 7 d. Conditions same as in Figure 4.1. (b) DCAA and TCAA.

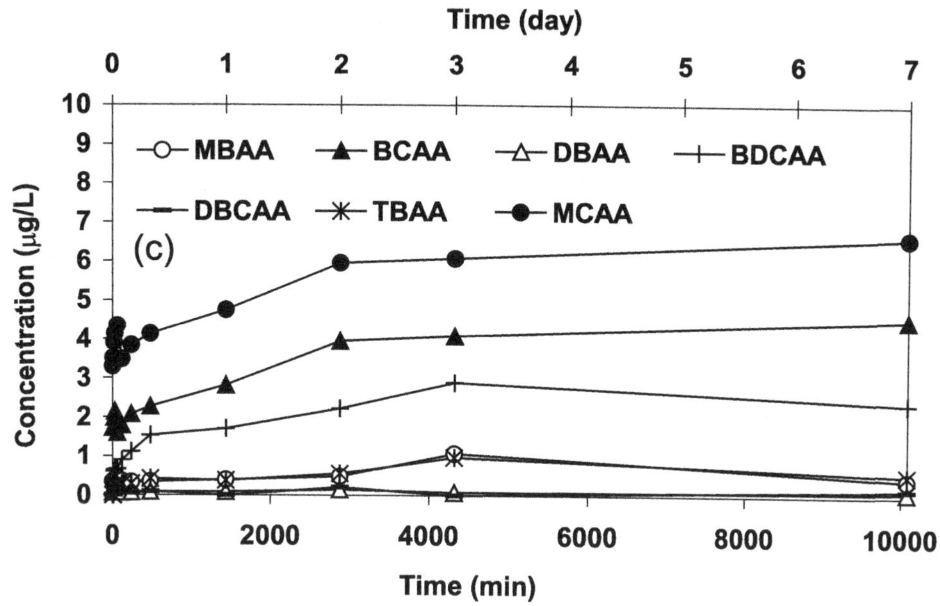

Figure 4.6c THM and HAA formation in LW water at reaction times from 5 min to 7 d. Conditions same as in Figure 4.1. (c) HAAs present at < 10 μg/L.

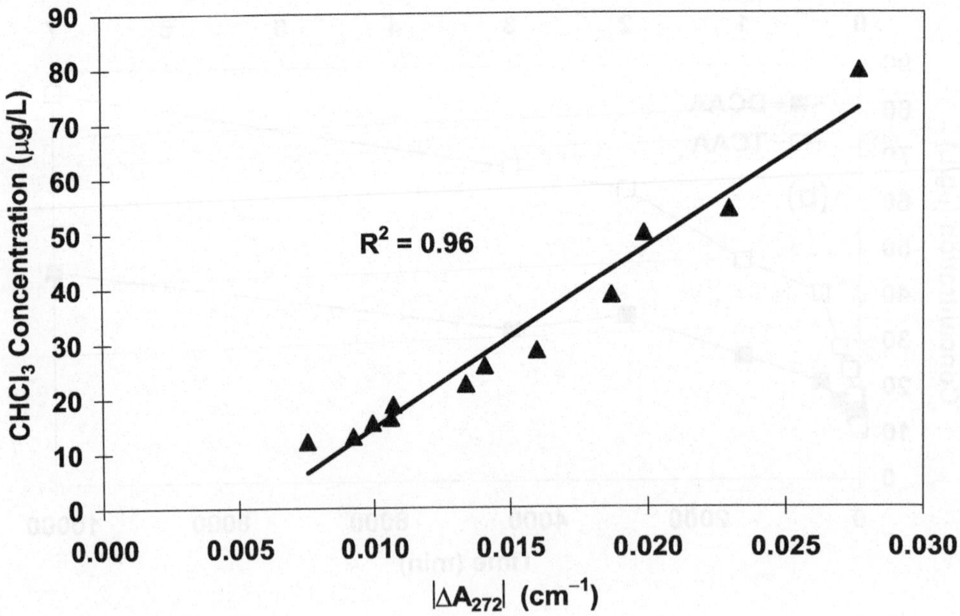

Figure 4.7 Correlation between $CHCl_3$ concentration and $|\Delta A_{272}|$ during chlorination of LW water. Conditions same as in Figure 4.1.

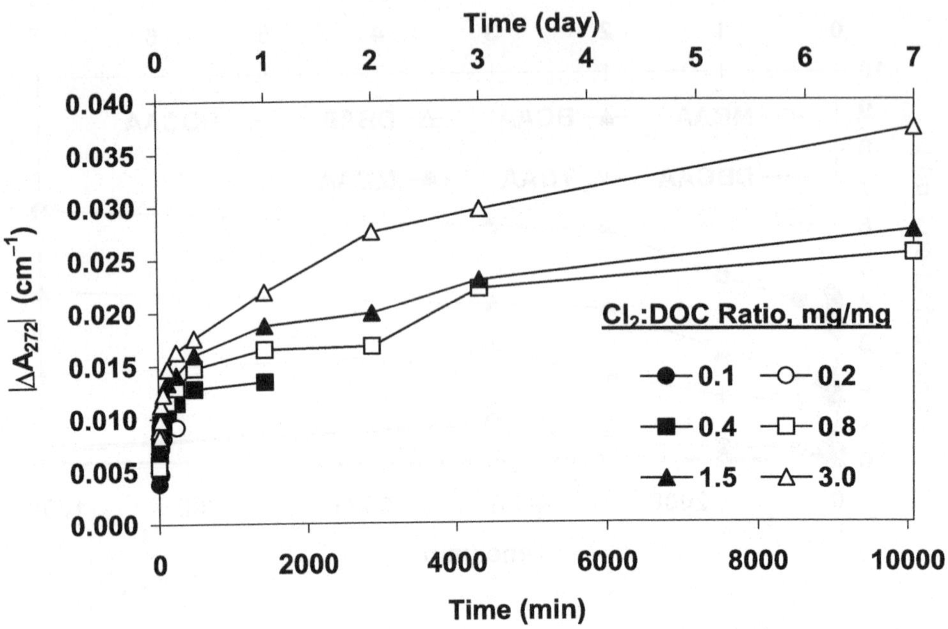

Figure 4.8 Development of $|\Delta A_{272}|$ over time during chlorination of LW water at various Cl_2:DOC ratios. pH 7, 25°C.

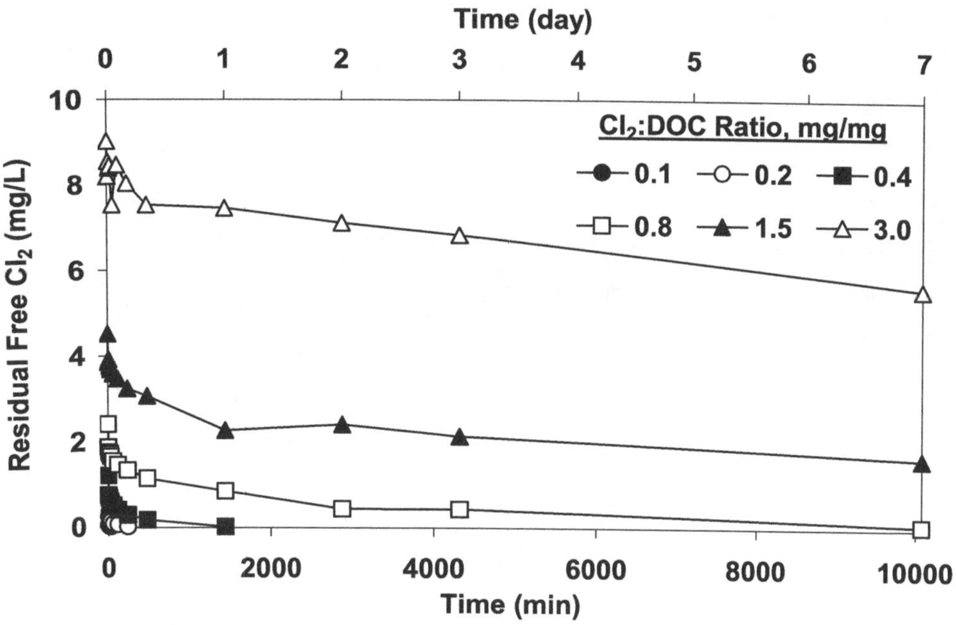

Figure 4.9 Depletion of free chlorine during chlorination of LW water at various Cl_2:DOC ratios. pH 7, 25°C.

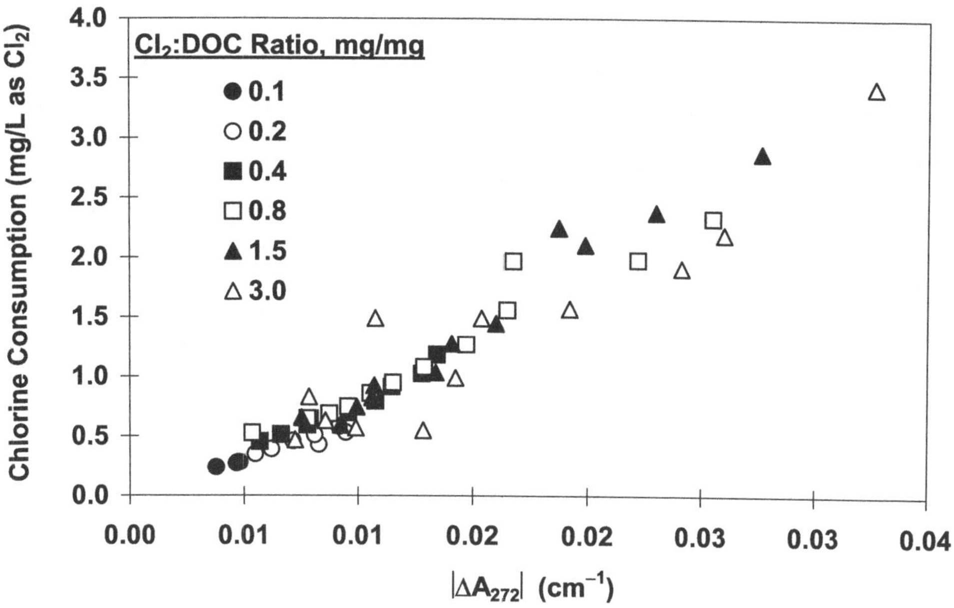

Figure 4.10 Correlation between chlorine consumption and $|\Delta A_{272}|$ for chlorination of LW water at various Cl_2:DOC ratios. pH 7, 25°C.

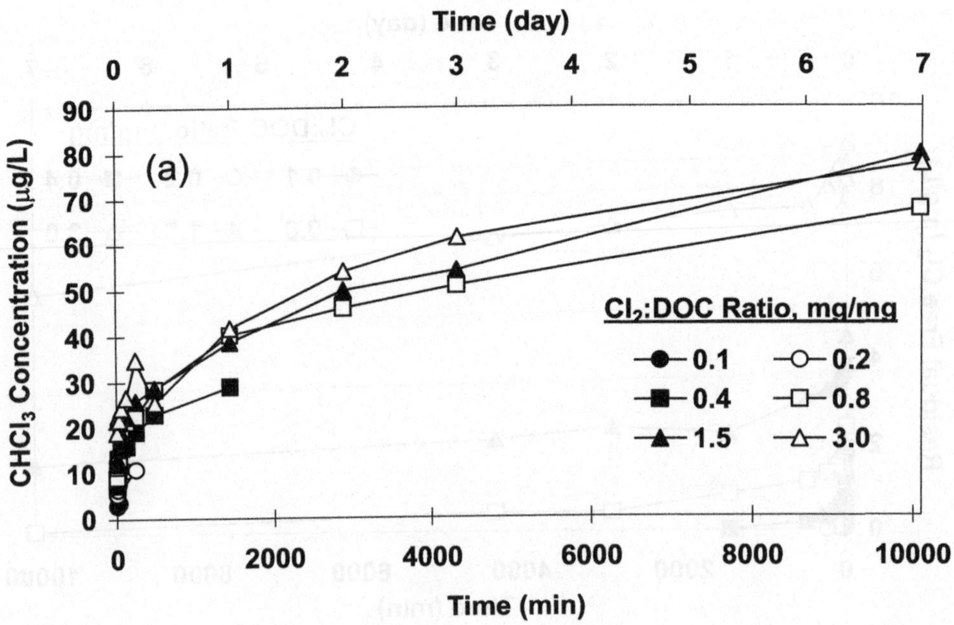

Figure 4.11a DBP formation kinetics during chlorination of LW water at various Cl_2:DOC ratios. pH 7, 25°C. (a) $CHCl_3$.

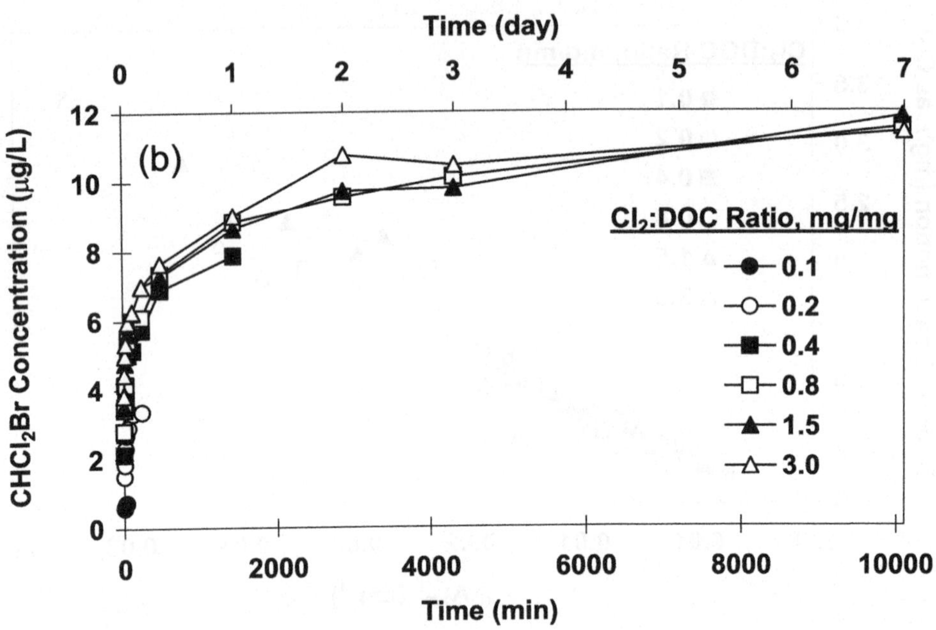

Figure 4.11b DBP formation kinetics during chlorination of LW water at various Cl_2:DOC ratios. pH 7, 25°C. (b) $CHCl_2Br$.

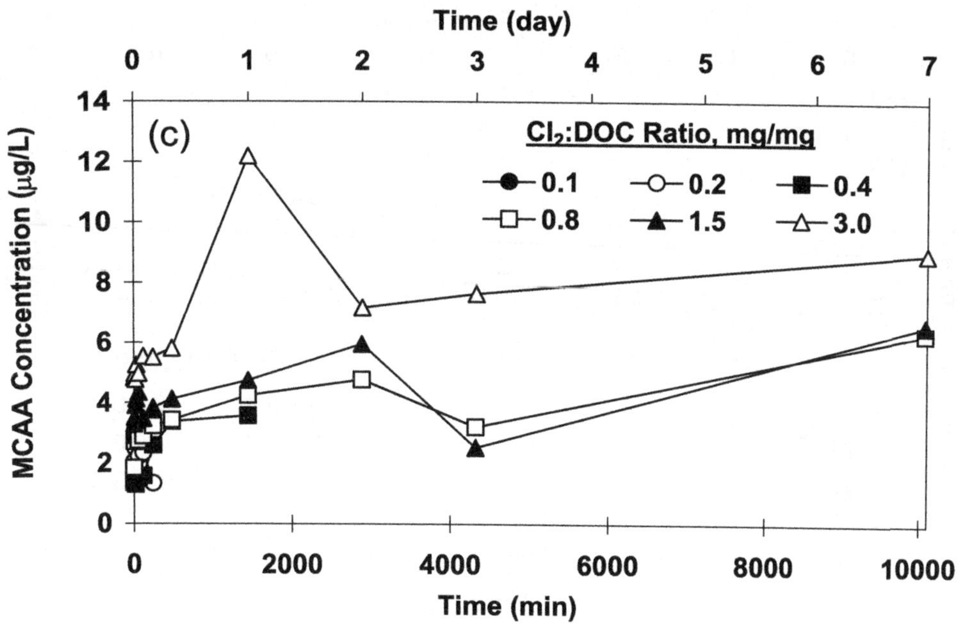

Figure 4.11c DBP formation kinetics during chlorination of LW water at various Cl_2:DOC ratios. pH 7, 25°C. (c) MCAA.

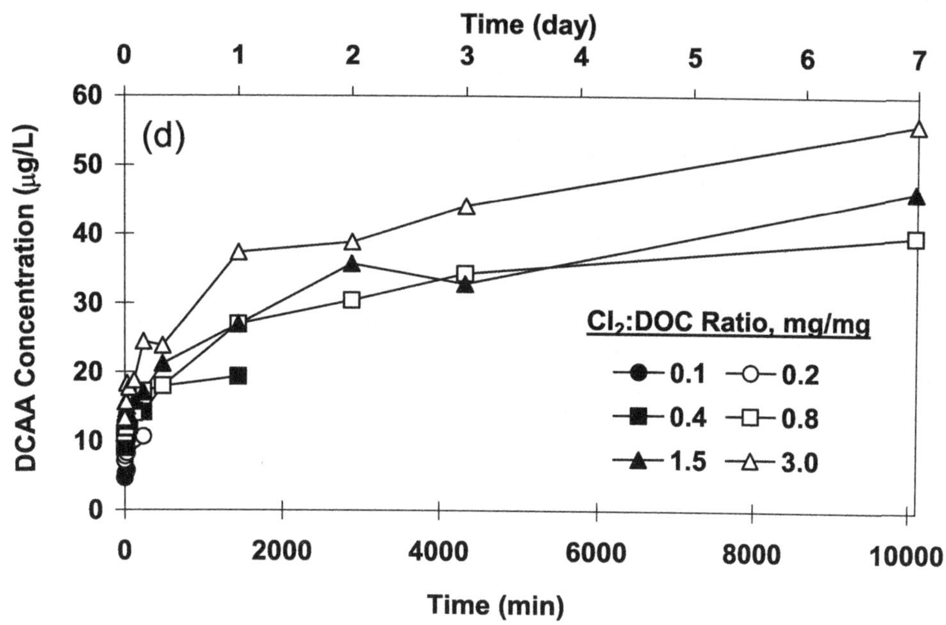

Figure 4.11d DBP formation kinetics during chlorination of LW water at various Cl_2:DOC ratios. pH 7, 25°C. (d) DCAA.

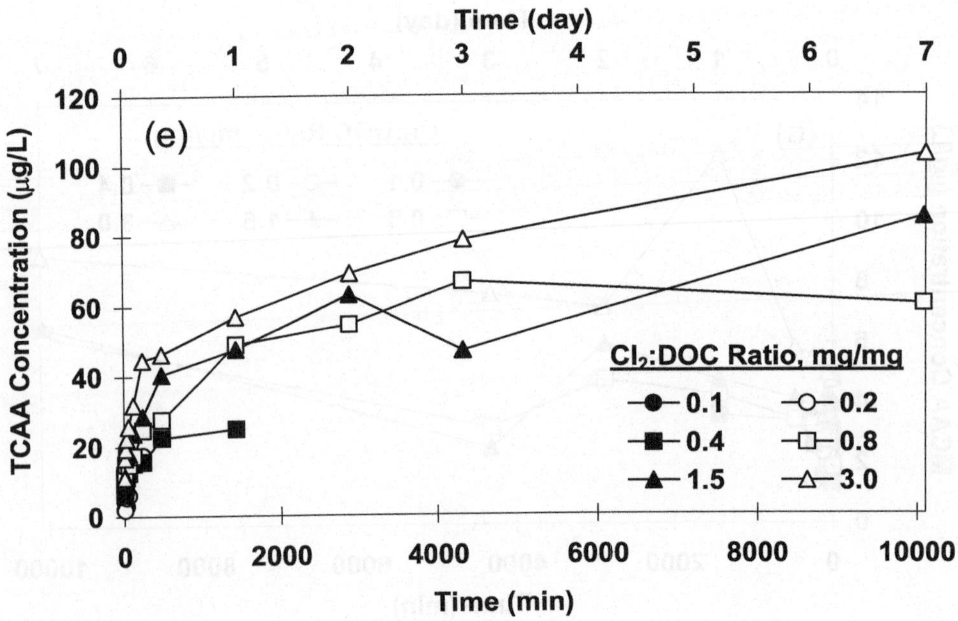

Figure 4.11e DBP formation kinetics during chlorination of LW water at various Cl_2:DOC ratios. pH 7, 25°C. (e) TCAA.

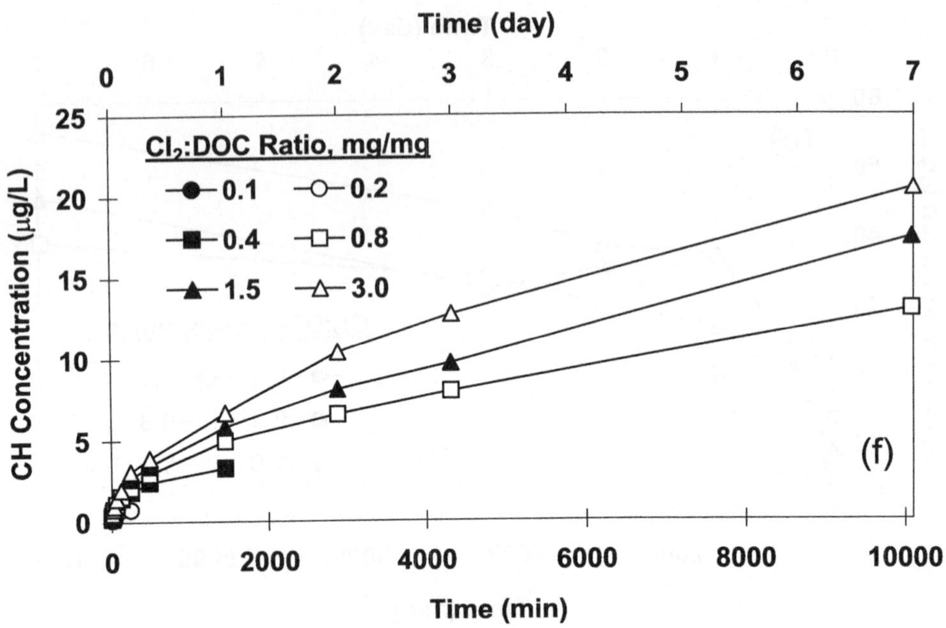

Figure 4.11f DBP formation kinetics during chlorination of LW water at various Cl_2:DOC ratios. pH 7, 25°C. (f) CH.

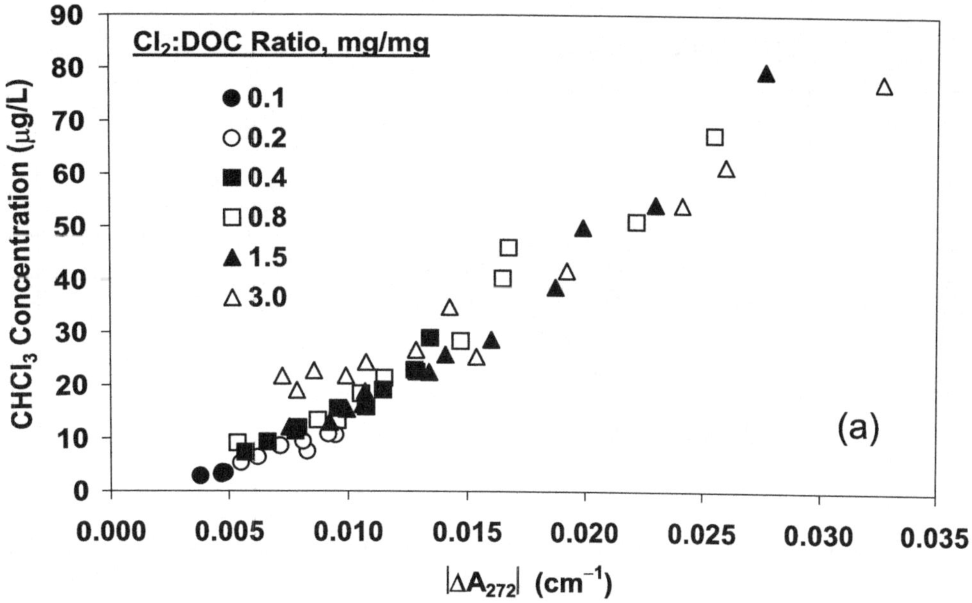

Figure 4.12a Correlation between DBP formation and $|\Delta A_{272}|$ during chlorination of LW water at various Cl_2:DOC ratios. pH 7, 25°C. (a) $CHCl_3$.

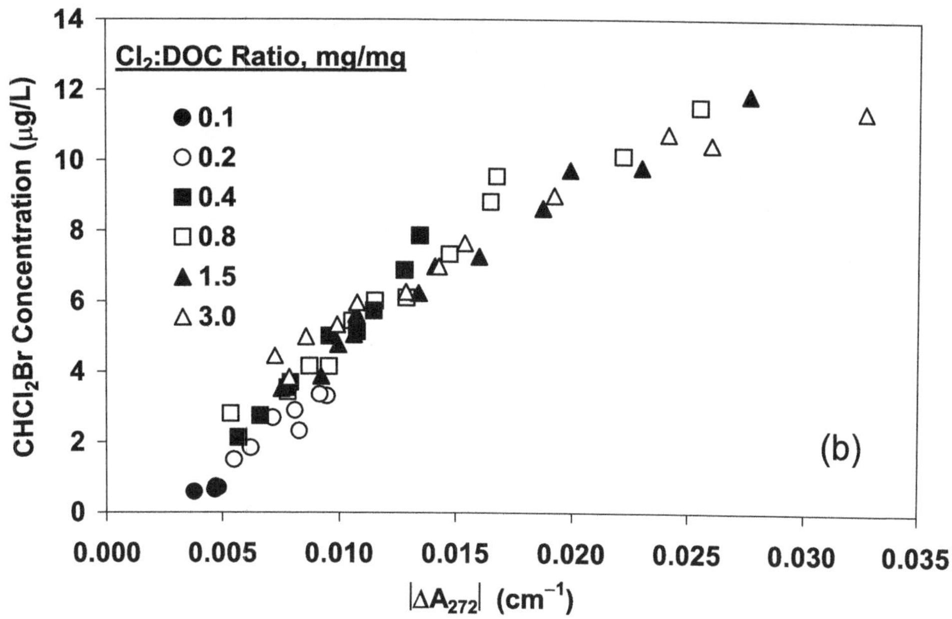

Figure 4.12b Correlation between DBP formation and $|\Delta A_{272}|$ during chlorination of LW water at various Cl_2:DOC ratios. pH 7, 25°C. (b) $CHCl_2Br$.

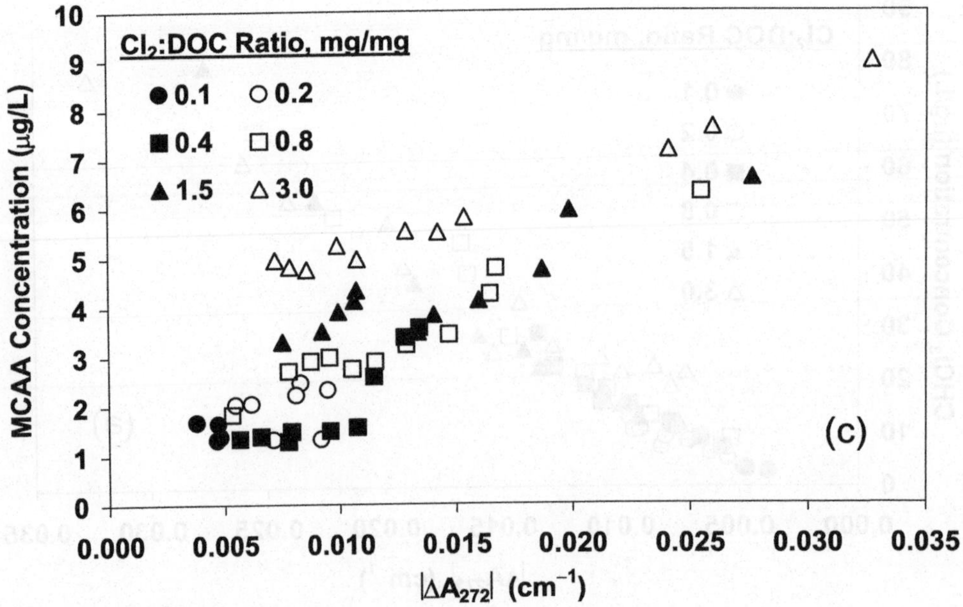

Figure 4.12c Correlation between DBP formation and $|\Delta A_{272}|$ during chlorination of LW water at various Cl_2:DOC ratios. pH 7, 25°C. (c) MCAA.

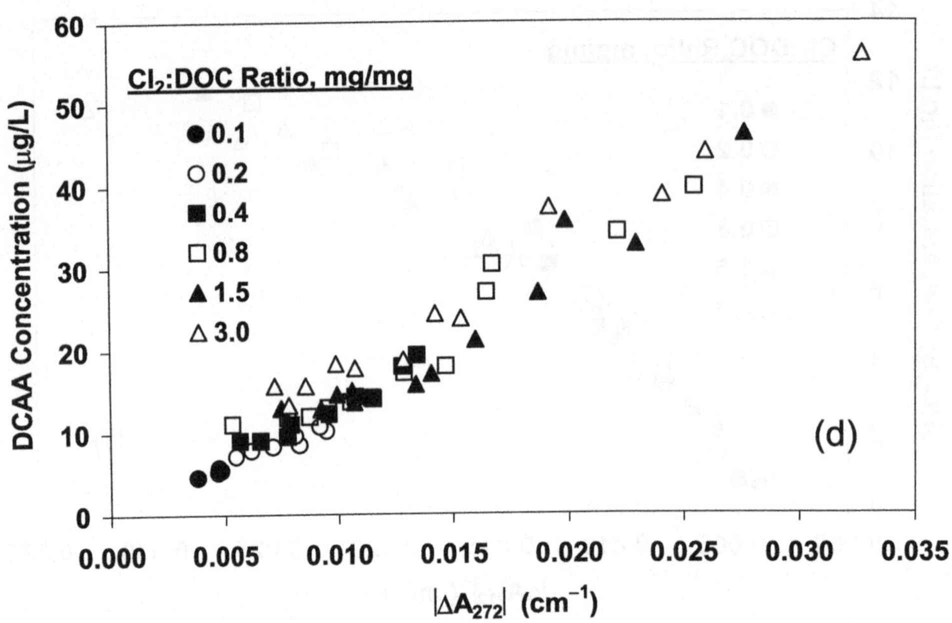

Figure 4.12d Correlation between DBP formation and $|\Delta A_{272}|$ during chlorination of LW water at various Cl_2:DOC ratios. pH 7, 25°C. (d) DCAA.

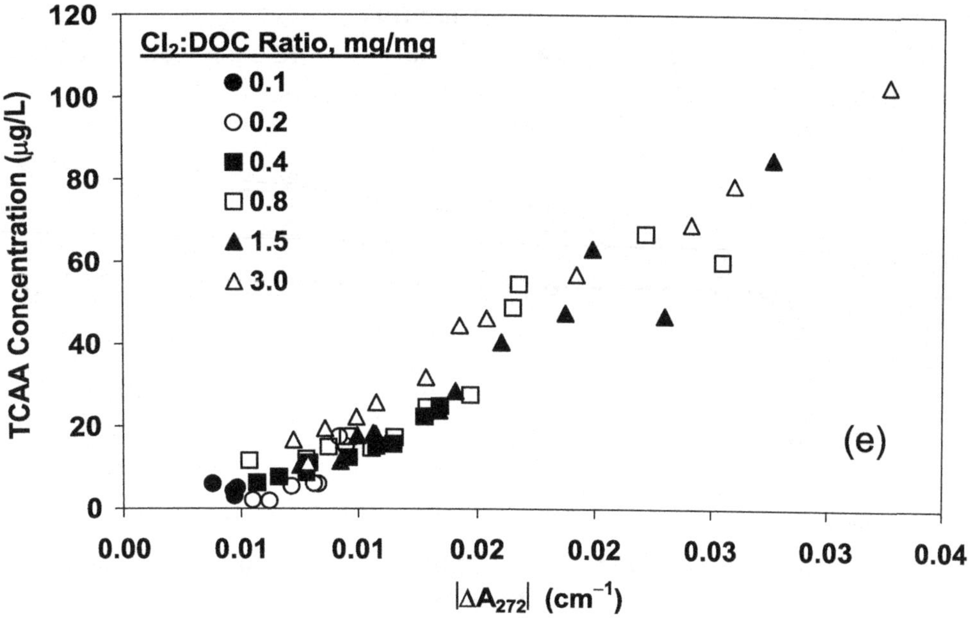

Figure 4.12e Correlation between DBP formation and $|\Delta A_{272}|$ during chlorination of LW water at various Cl_2:DOC ratios. pH 7, 25°C. (e) TCAA.

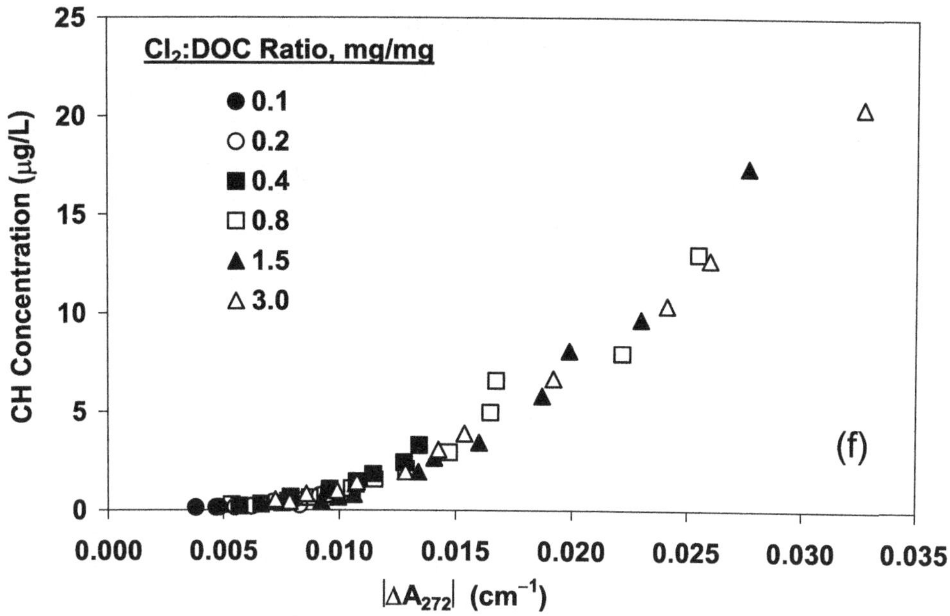

Figure 4.12f Correlation between DBP formation and $|\Delta A_{272}|$ during chlorination of LW water at various Cl_2:DOC ratios. pH 7, 25°C. (f) CH.

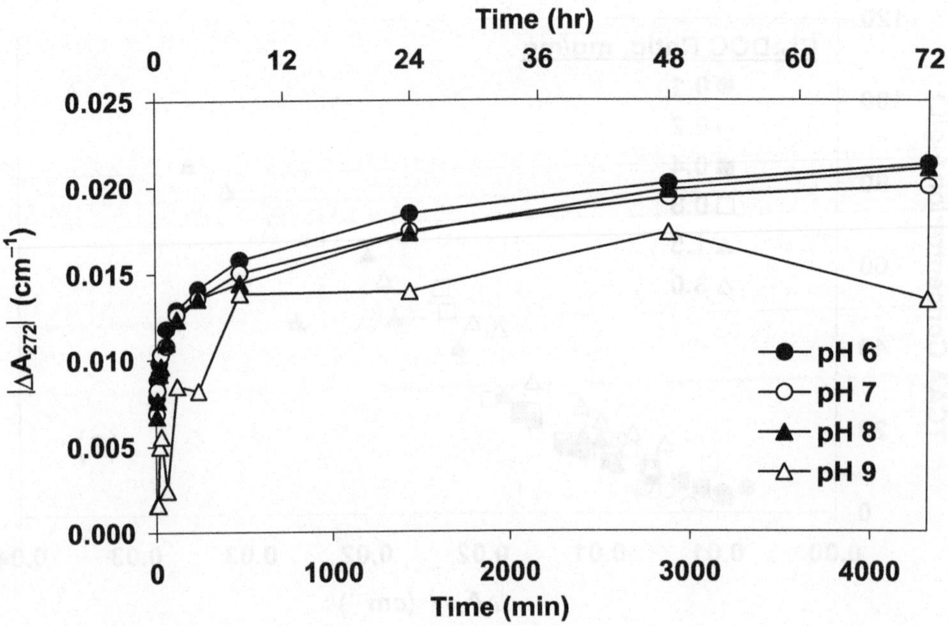

Figure 4.13 Development of $|\Delta A_{272}|$ for chlorination of LW water at a range of pH values. Cl_2:DOC ratio 1.5, pH 7, 25°C.

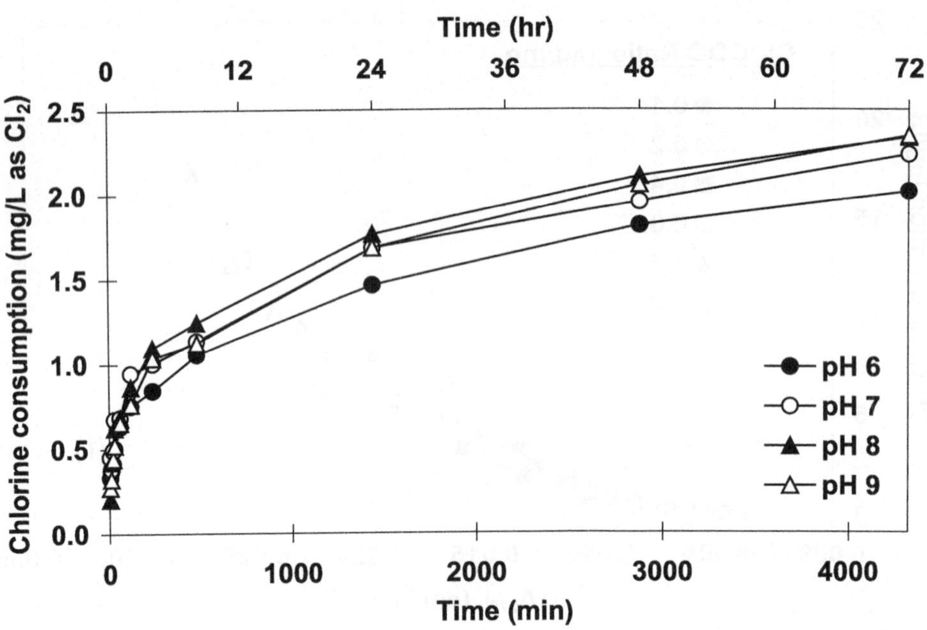

Figure 4.14 Chlorine consumption during chlorination of LW water at a range of pH values. Cl_2:DOC ratio 1.5, pH 7, 25°C.

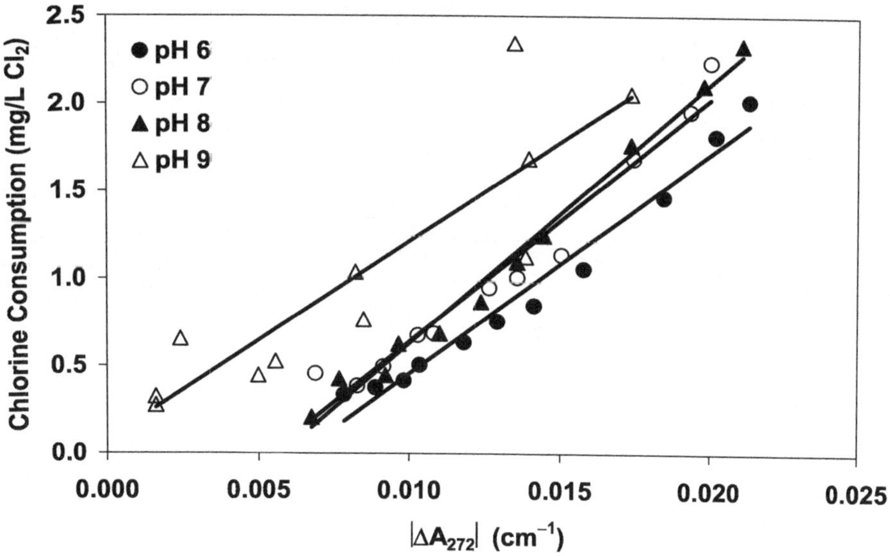

Figure 4.15 Relationship between chlorine consumption and $|\Delta A_{272}|$ for chlorination of LW water at various pH's. Cl_2:DOC 1.5, pH 7, 25°C.

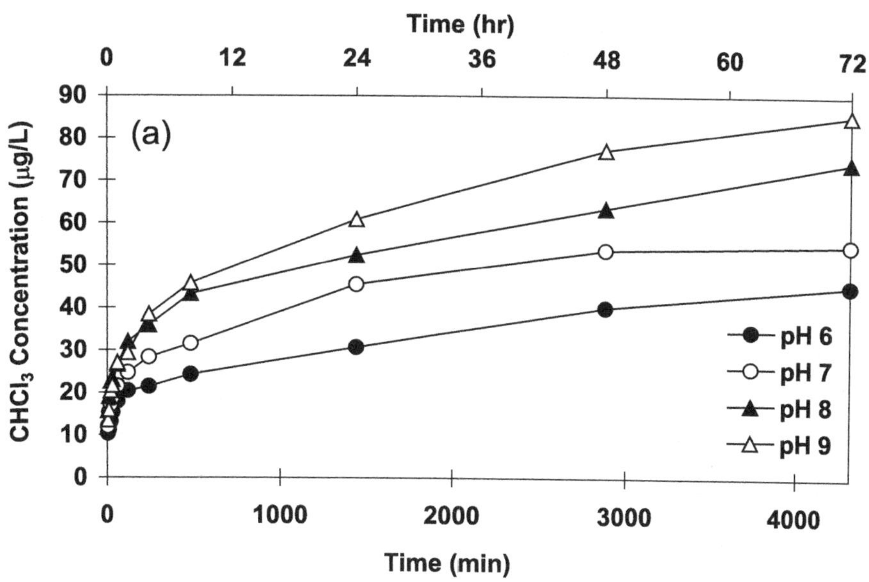

Figure 4.16a DBP formation from chlorination of LW water at various pH values. Cl_2:DOC ratio 1.5, 25°C. (a) $CHCl_3$.

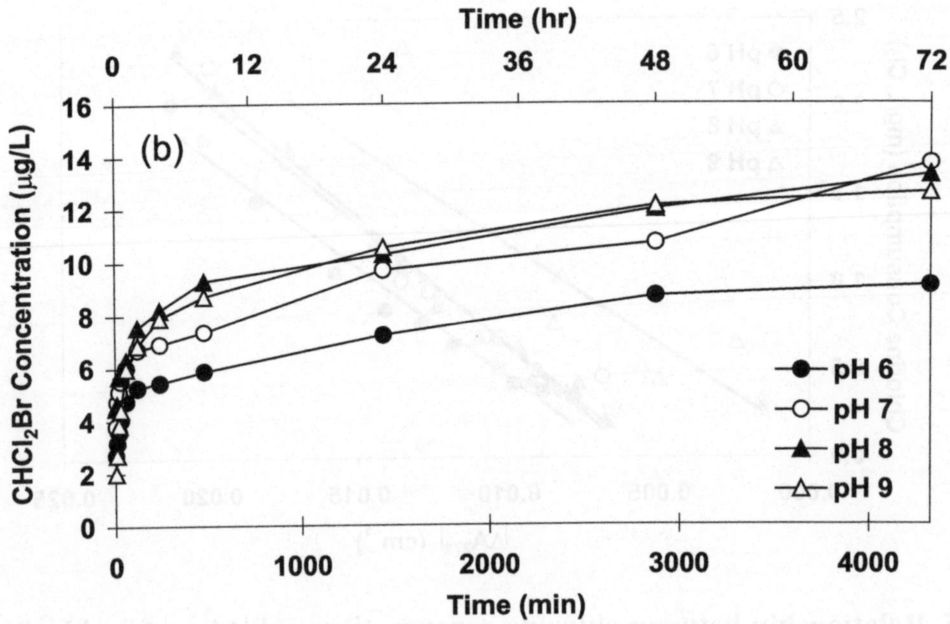

Figure 4.16b DBP formation from chlorination of LW water at various pH values. Cl_2:DOC ratio 1.5, 25°C. (b) $CHCl_2Br$.

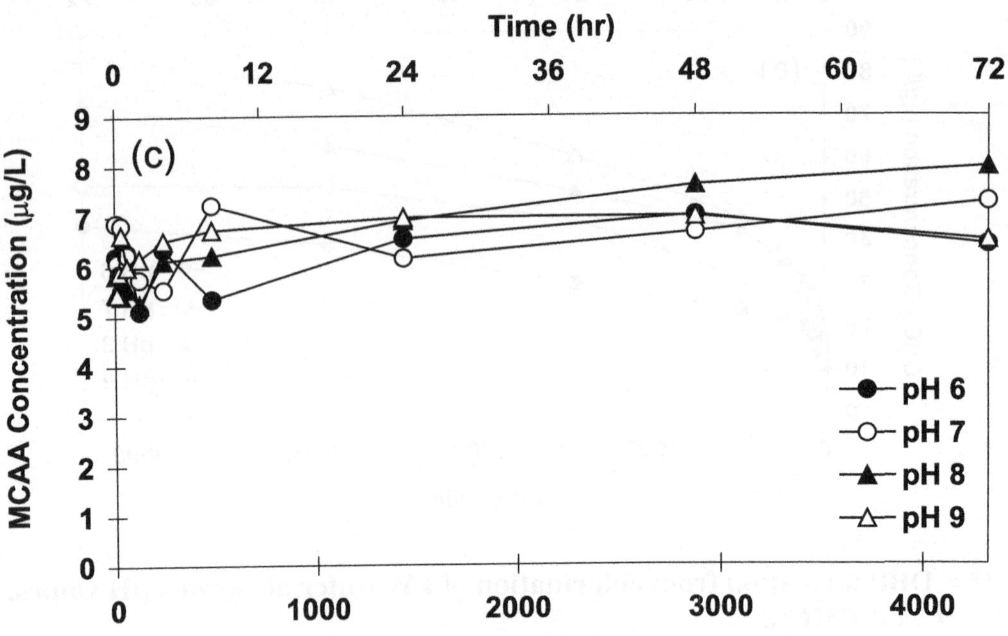

Figure 4.16c DBP formation from chlorination of LW water at various pH values. Cl_2:DOC ratio 1.5, 25°C. (c) MCAA.

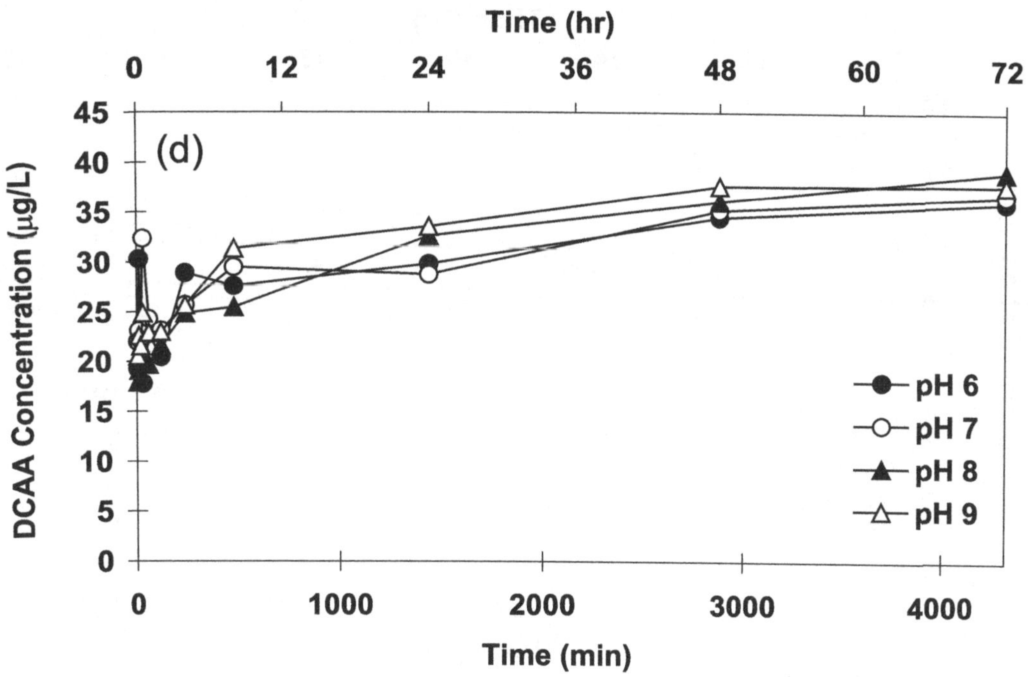

Figure 4.16d **DBP formation from chlorination of LW water at various pH values. Cl_2:DOC ratio 1.5, 25°C. (d) DCAA.**

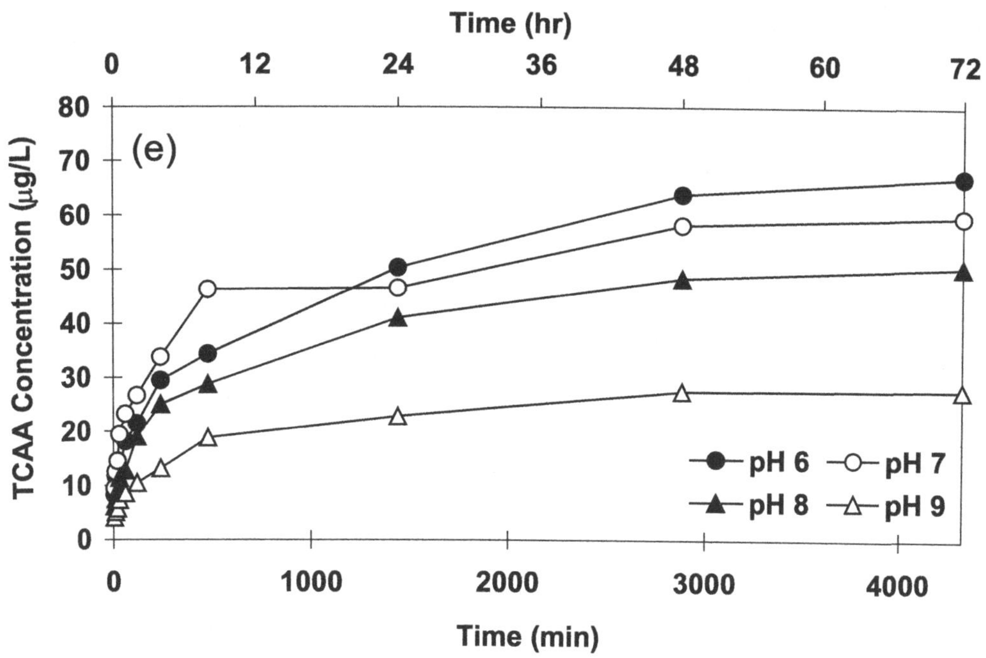

Figure 4.16e **DBP formation from chlorination of LW water at various pH values. Cl_2:DOC ratio 1.5, 25°C. (e) TCAA.**

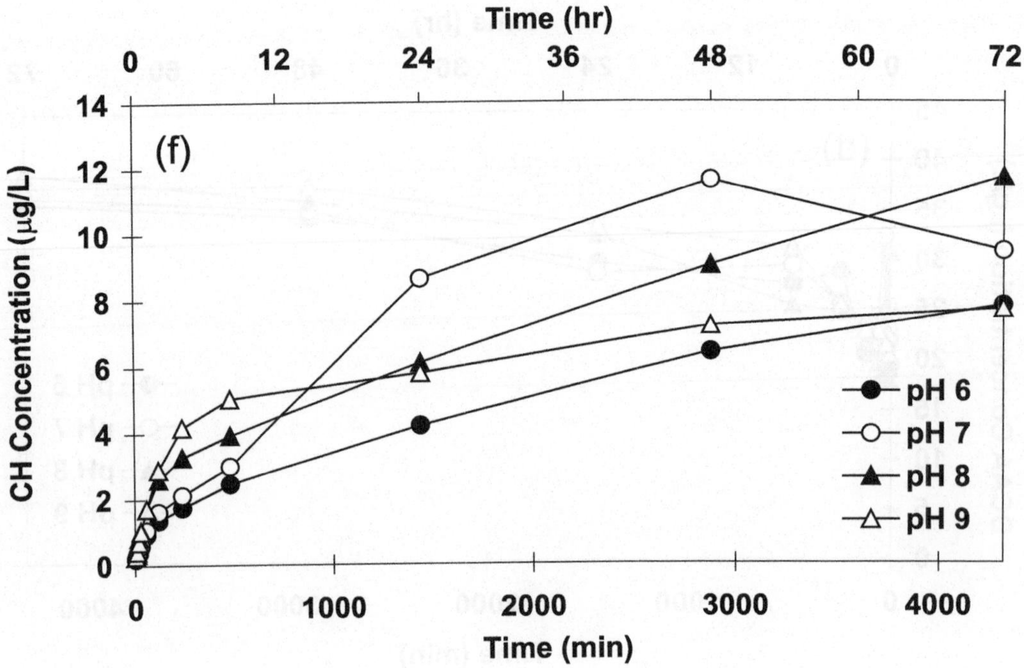

Figure 4.16f DBP formation from chlorination of LW water at various pH values. Cl_2:DOC ratio 1.5, 25°C. (f) CH.

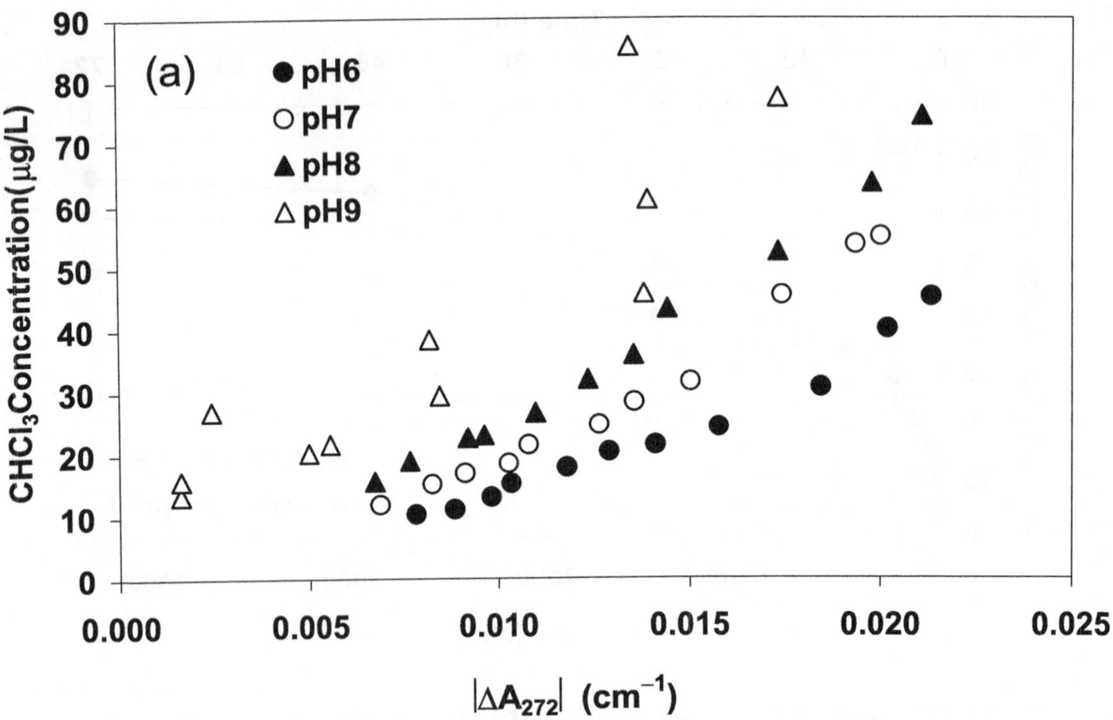

Figure 4.17a DBP formation as a function of $|\Delta A_{272}|$ for chlorination of LW water at various pH's. Cl_2:DOC ratio 1.5, pH 7, 25°C. (a) $CHCl_3$.

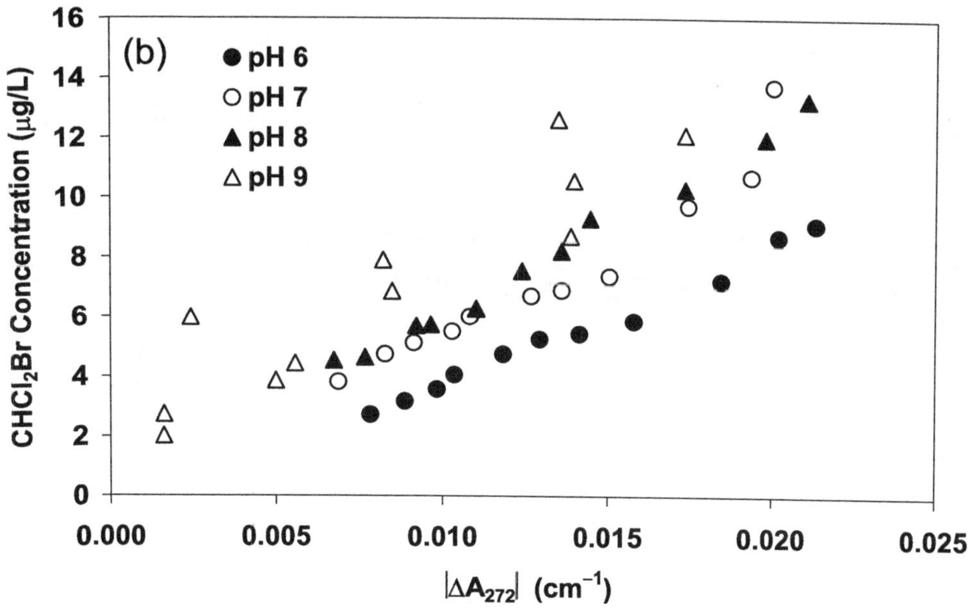

Figure 4.17b DBP formation as a function of $|\Delta A_{272}|$ for chlorination of LW water at various pH's. Cl_2:DOC ratio 1.5, pH 7, 25°C. (b) $CHCl_2Br$.

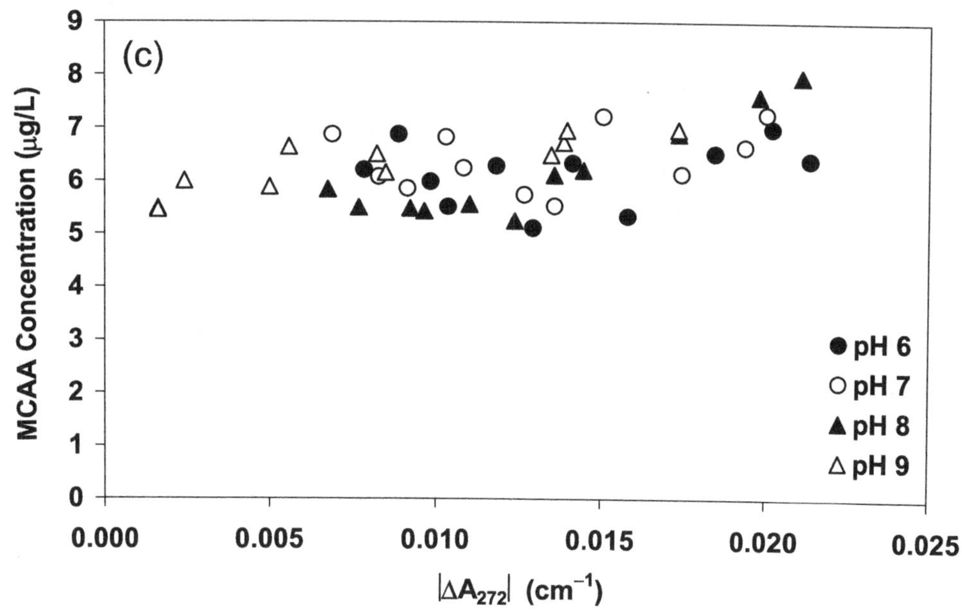

Figure 4.17c DBP formation as a function of $|\Delta A_{272}|$ for chlorination of LW water at various pH's. Cl_2:DOC ratio 1.5, pH 7, 25°C. (c) MCAA.

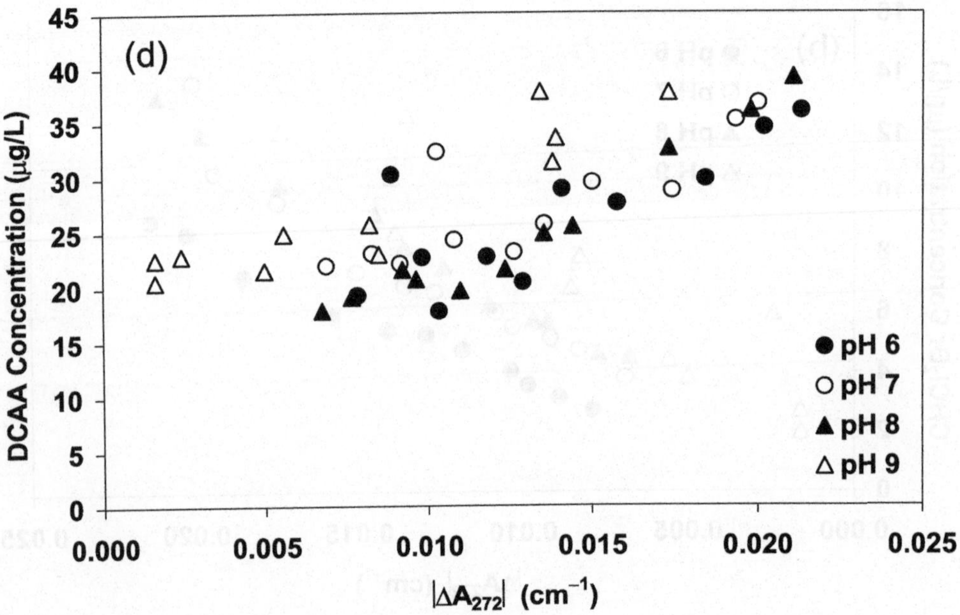

Figure 4.17d DBP formation as a function of $|\Delta A_{272}|$ for chlorination of LW water at various pH's. Cl_2:DOC ratio 1.5, pH 7, 25°C. (d) DCAA.

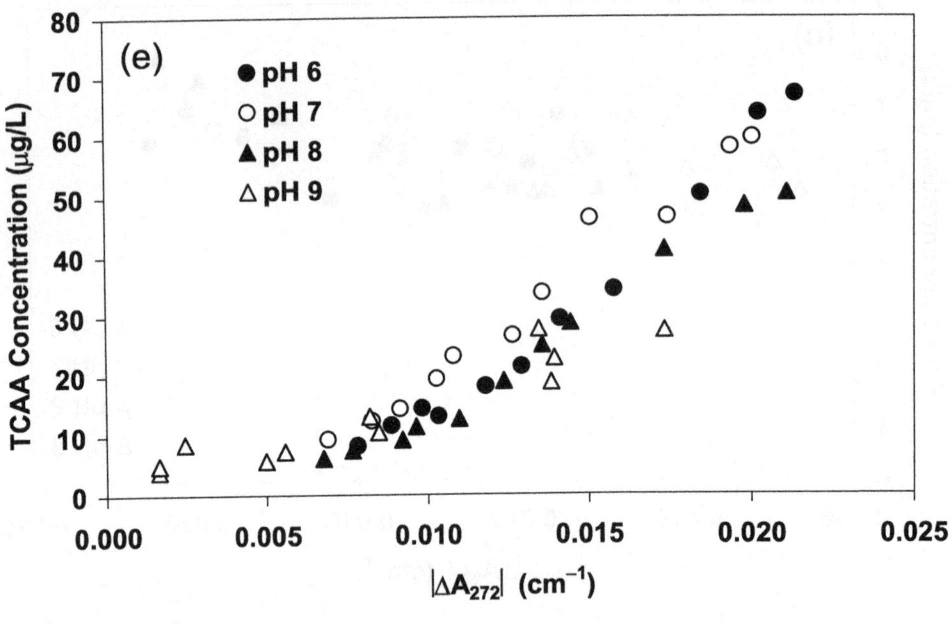

Figure 4.17e DBP formation as a function of $|\Delta A_{272}|$ for chlorination of LW water at various pH's. Cl_2:DOC ratio 1.5, pH 7, 25°C. (e) TCAA.

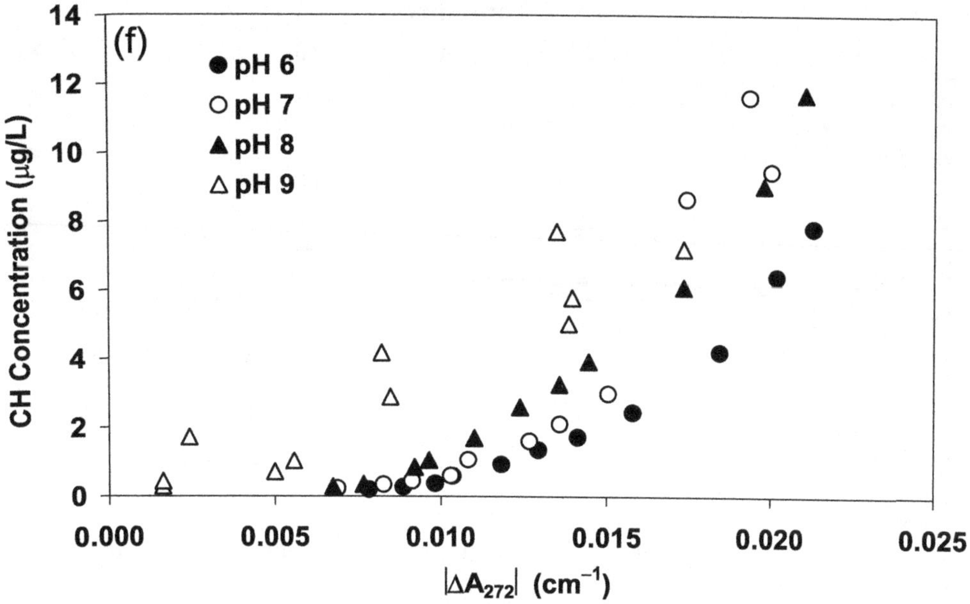

Figure 4.17f DBP formation as a function of $|\Delta A_{272}|$ for chlorination of LW water at various pH's. Cl_2:DOC ratio 1.5, pH 7, 25°C. (f) CH.

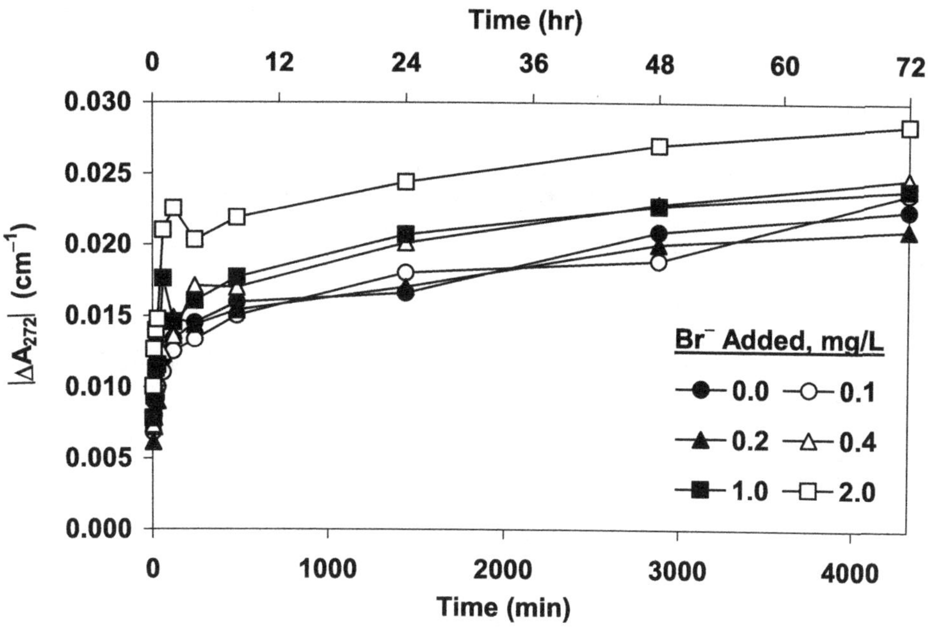

Figure 4.18 Development of $|\Delta A_{272}|$ during chlorination of LW water amended with varying concentrations of Br^-. Cl_2:DOC ratio 1.5, pH 7, 25°C.

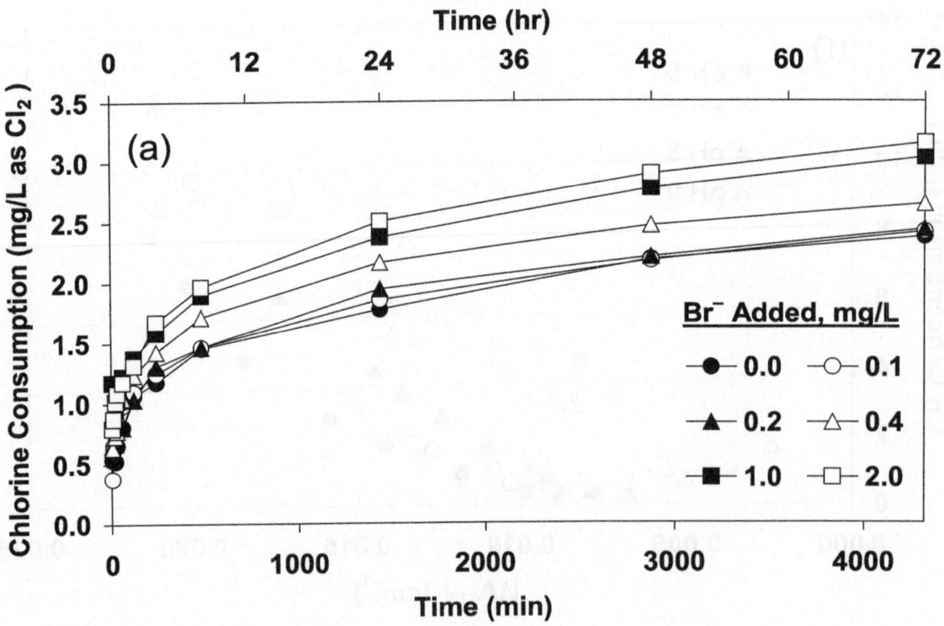

Figure 4.19a Kinetics of chlorine consumption in systems with varying Br^- concentrations. LW water, Cl_2:DOC ratio 1.5, pH 7, 25°C. (a) Trends over 72 hr.

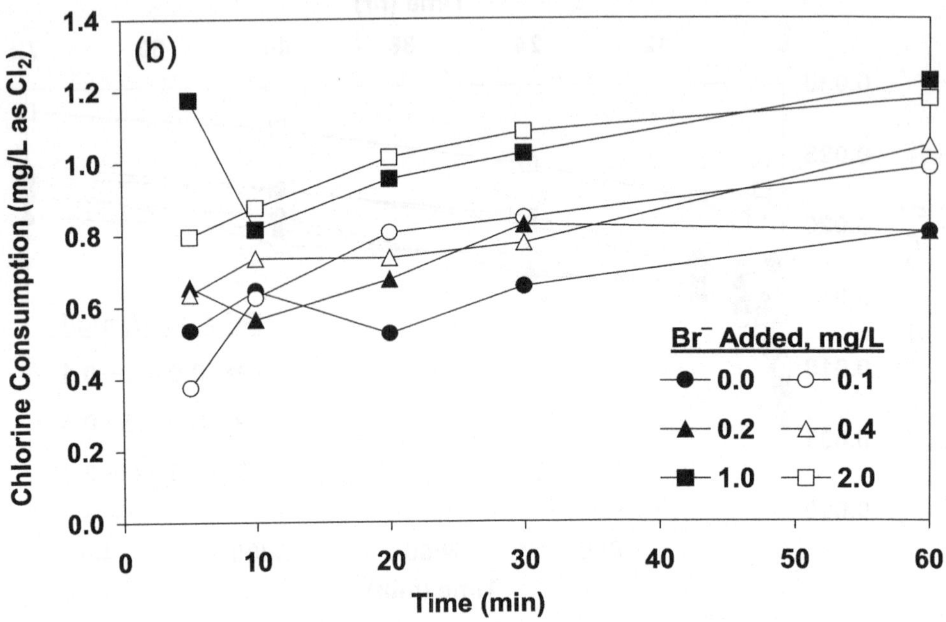

Figure 4.19b Kinetics of chlorine consumption in systems with varying Br^- concentrations. LW water, Cl_2:DOC ratio 1.5, pH 7, 25°C. (b) Trends during initial 60 min.

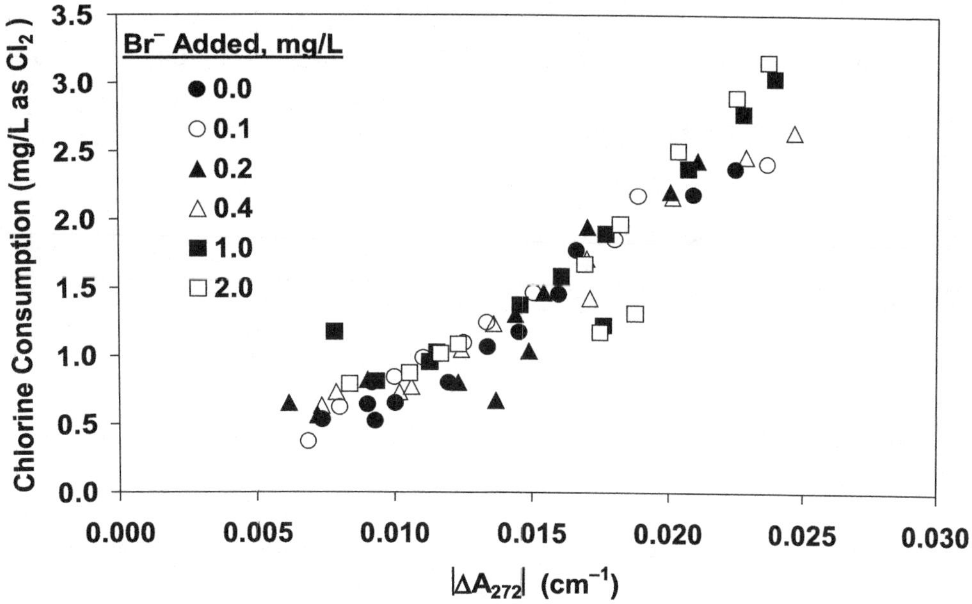

Figure 4.20 Relationship between chlorine consumption and $|\Delta A_{272}|$ in systems with varying Br^- concentrations. LW water, Cl_2:DOC ratio 1.5, pH 7, 25°C.

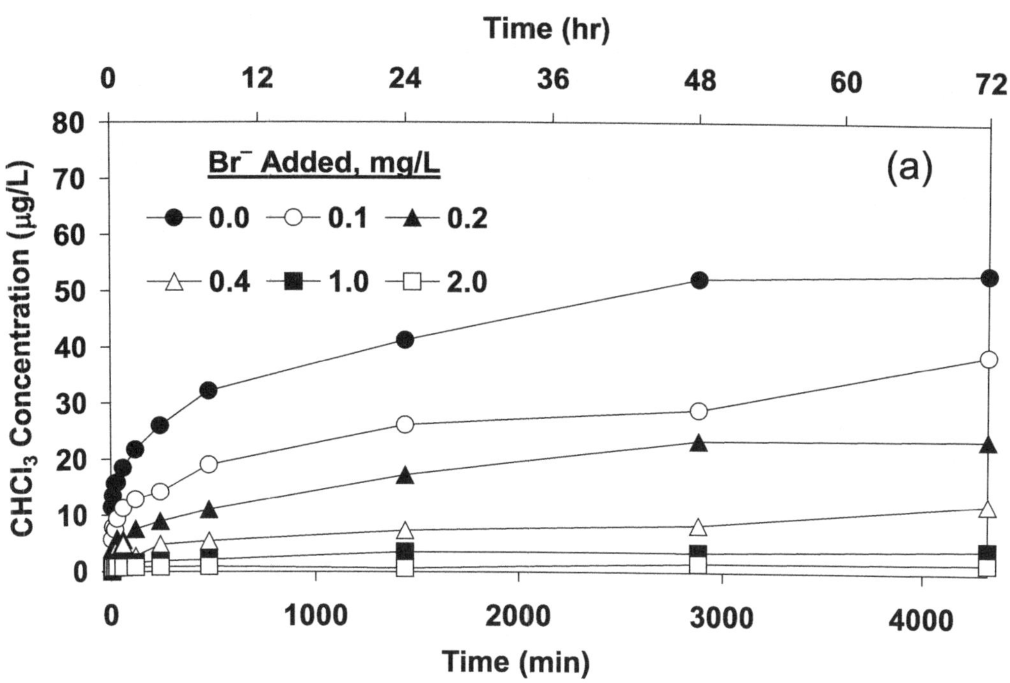

Figure 4.21a DBP formation kinetics in solutions with various Br^- additions. LW water, Cl_2:DOC ratio 1.5, pH 7, 25°C. (a) $CHCl_3$.

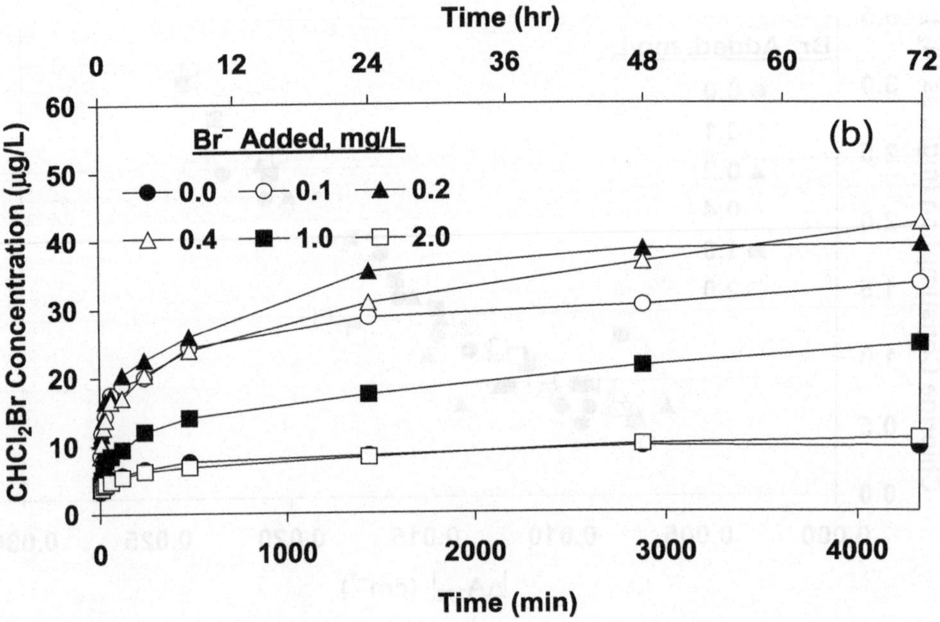

Figure 4.21b DBP formation kinetics in solutions with various Br^- additions. LW water, Cl_2:DOC ratio 1.5, pH 7, 25°C. (b) $CHCl_2Br$.

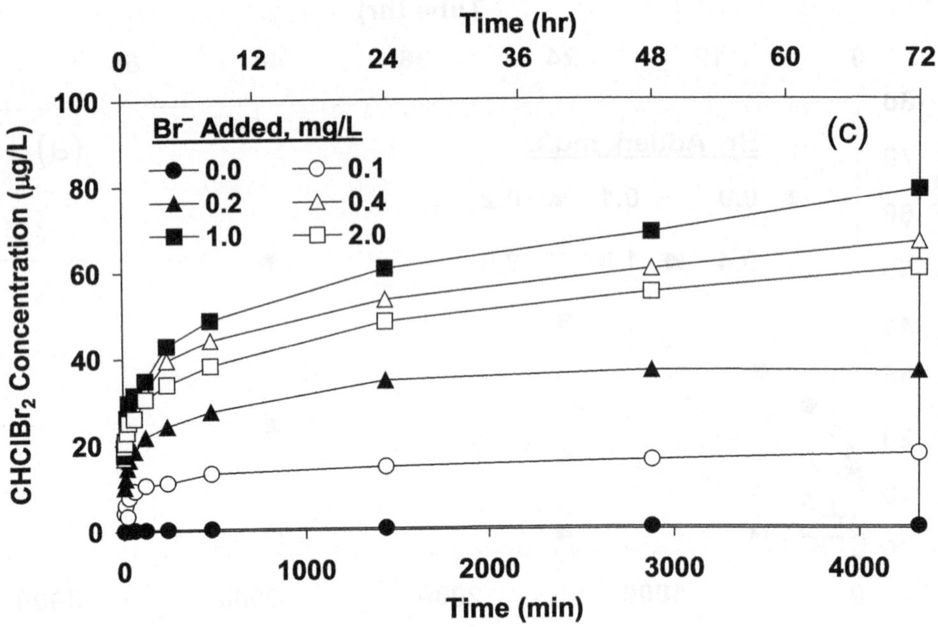

Figure 4.21c DBP formation kinetics in solutions with various Br^- additions. LW water, Cl_2:DOC ratio 1.5, pH 7, 25°C. (c) $CHClBr_2$.

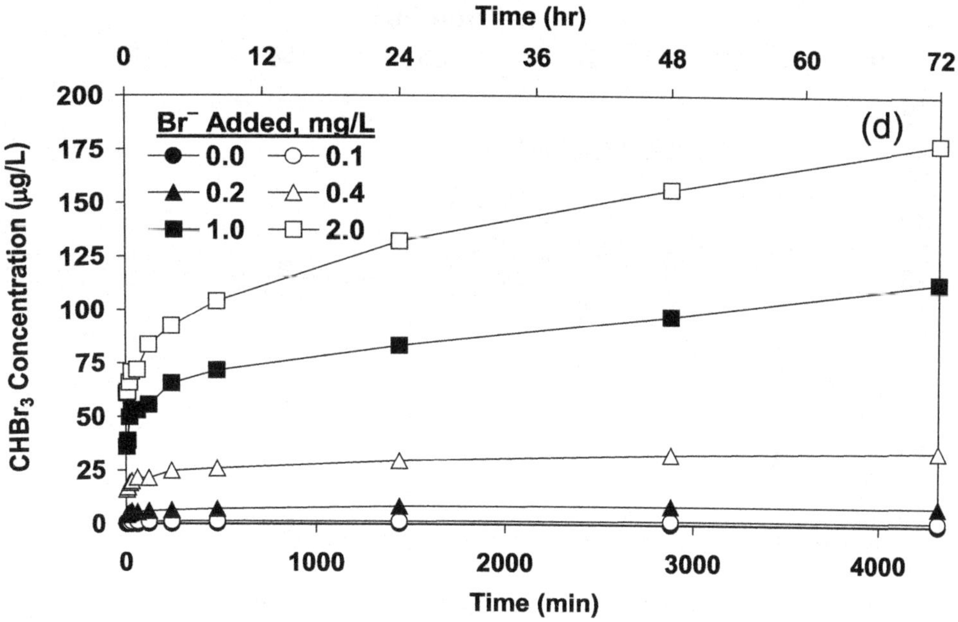

Figure 4.21d DBP formation kinetics in solutions with various Br^- additions. LW water, Cl_2:DOC ratio 1.5, pH 7, 25°C. (d) $CHBr_3$.

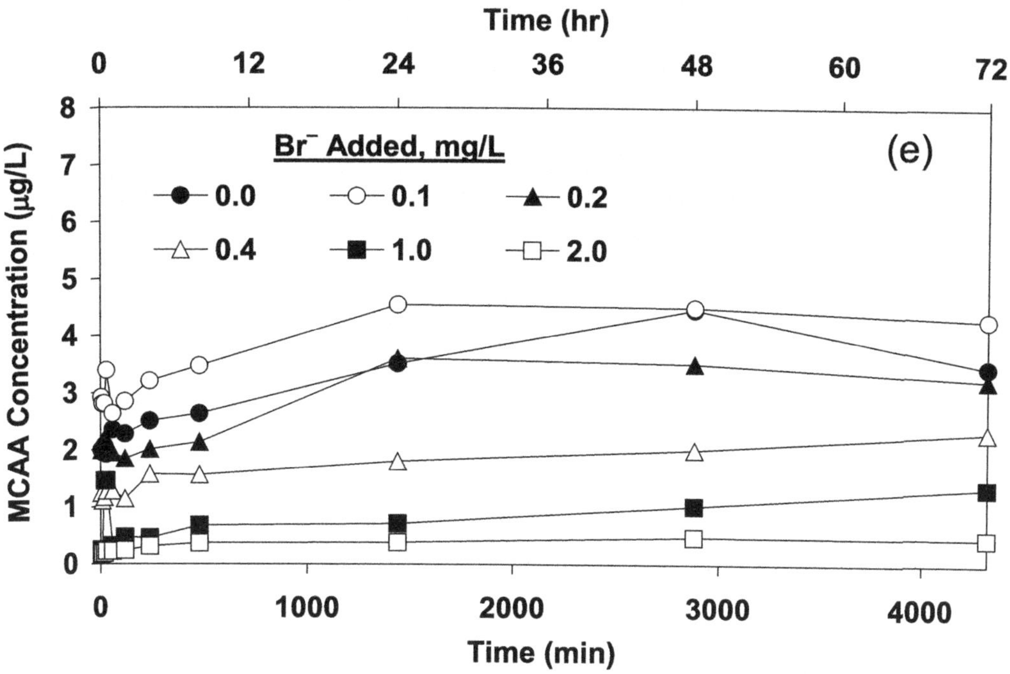

Figure 4.21e DBP formation kinetics in solutions with various Br^- additions. LW water, Cl_2:DOC ratio 1.5, pH 7, 25°C. (e) MCAA.

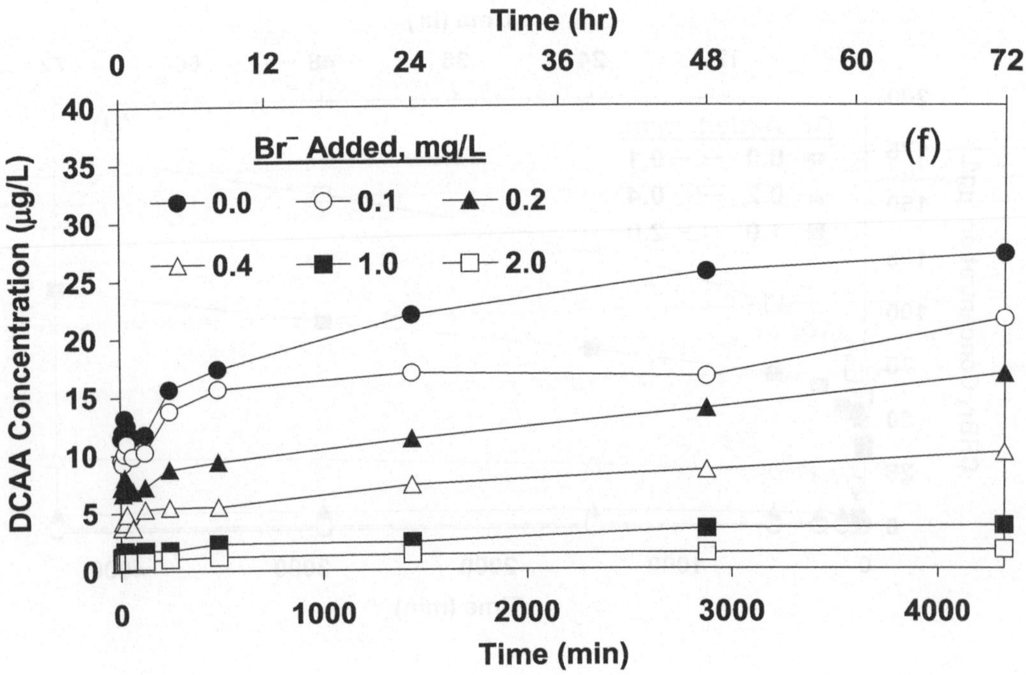

Figure 4.21f DBP formation kinetics in solutions with various Br⁻ additions. LW water, Cl_2:DOC ratio 1.5, pH 7, 25°C. (f) DCAA.

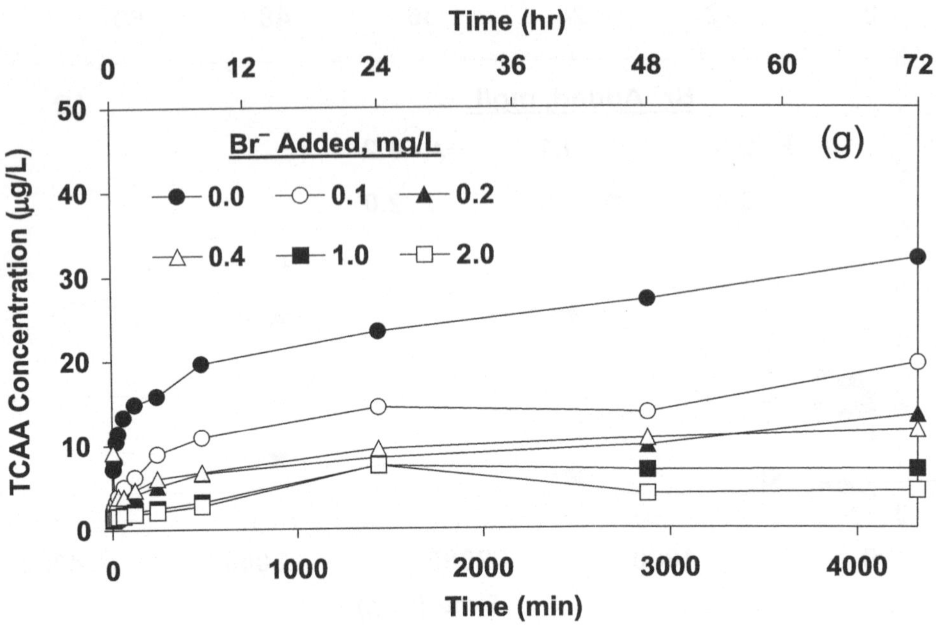

Figure 4.21g DBP formation kinetics in solutions with various Br⁻ additions. LW water, Cl_2:DOC ratio 1.5, pH 7, 25°C. (g) TCAA.

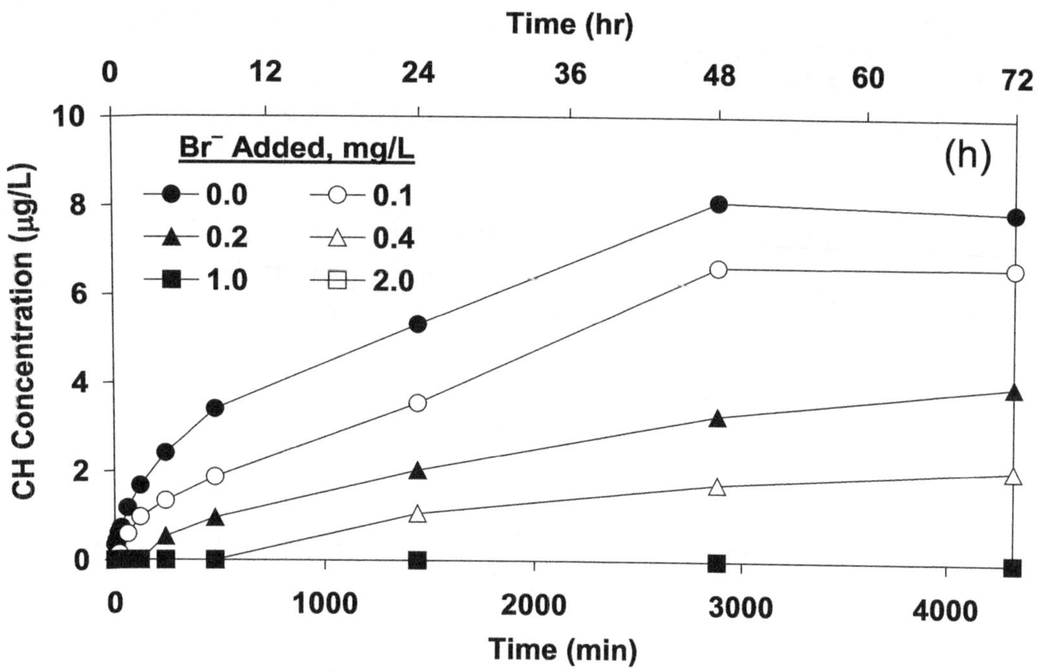

Figure 4.21h DBP formation kinetics in solutions with various Br^- additions. LW water, Cl_2:DOC ratio 1.5, pH 7, 25°C. (h) CH.

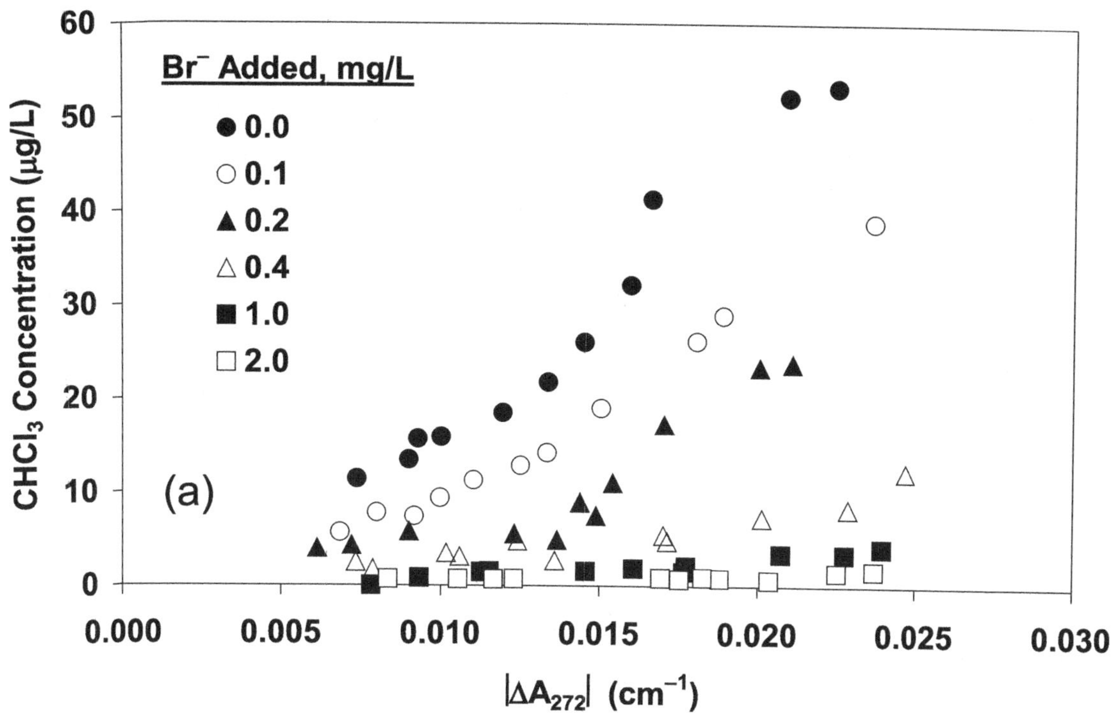

Figure 4.22a Relationship between DBP formation and $|\Delta A_{272}|$ in systems with varying Br^- concentration. LW water, Cl_2:DOC ratio 1.5, pH 7, 25°C. (a) $CHCl_3$.

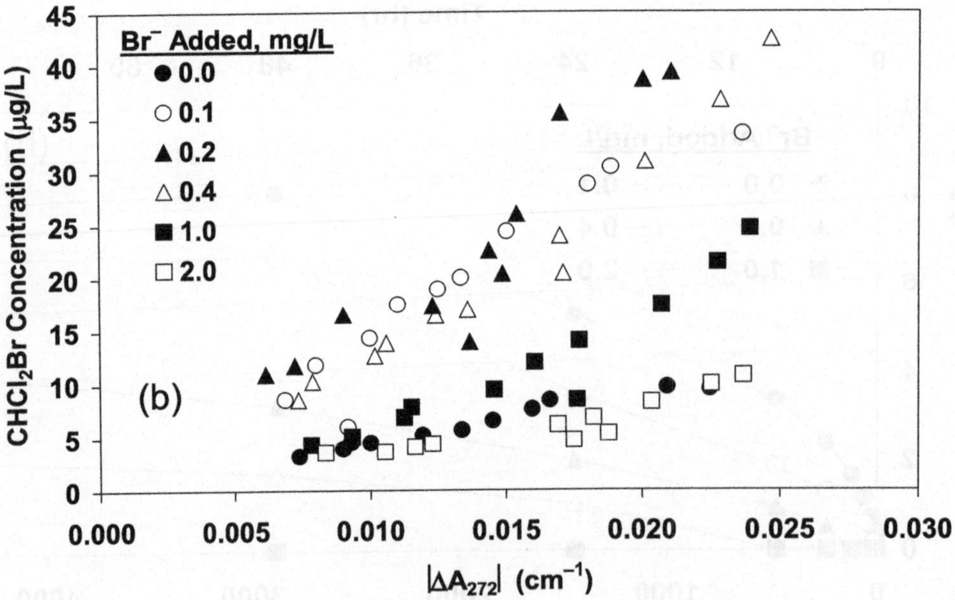

Figure 4.22b Relationship between DBP formation and $|\Delta A_{272}|$ in systems with varying Br^- concentration. LW water, Cl_2:DOC ratio 1.5, pH 7, 25°C. (b) $CHCl_2Br$.

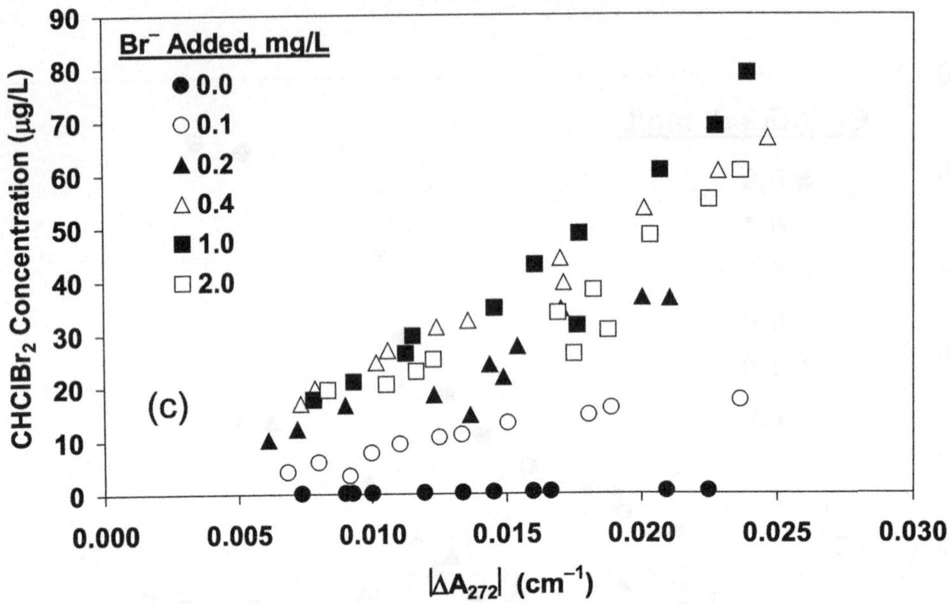

Figure 4.22c Relationship between DBP formation and $|\Delta A_{272}|$ in systems with varying Br^- concentration. LW water, Cl_2:DOC ratio 1.5, pH 7, 25°C. (c) $CHClBr_2$.

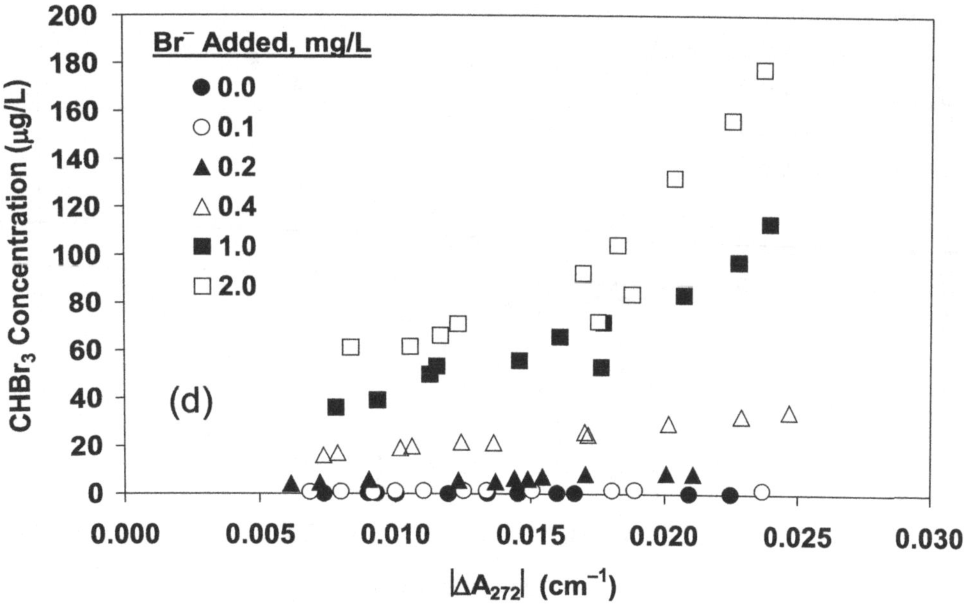

Figure 4.22d Relationship between DBP formation and $|\Delta A_{272}|$ in systems with varying Br^- concentration. LW water, Cl_2:DOC ratio 1.5, pH 7, 25°C. (d) $CHBr_3$.

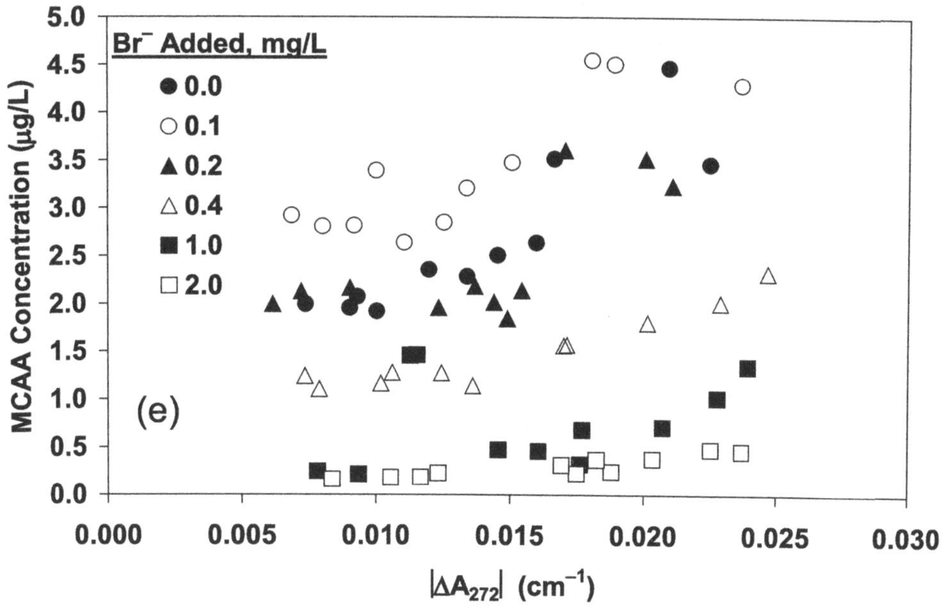

Figure 4.22e Relationship between DBP formation and $|\Delta A_{272}|$ in systems with varying Br^- concentration. LW water, Cl_2:DOC ratio 1.5, pH 7, 25°C. (e) MCAA.

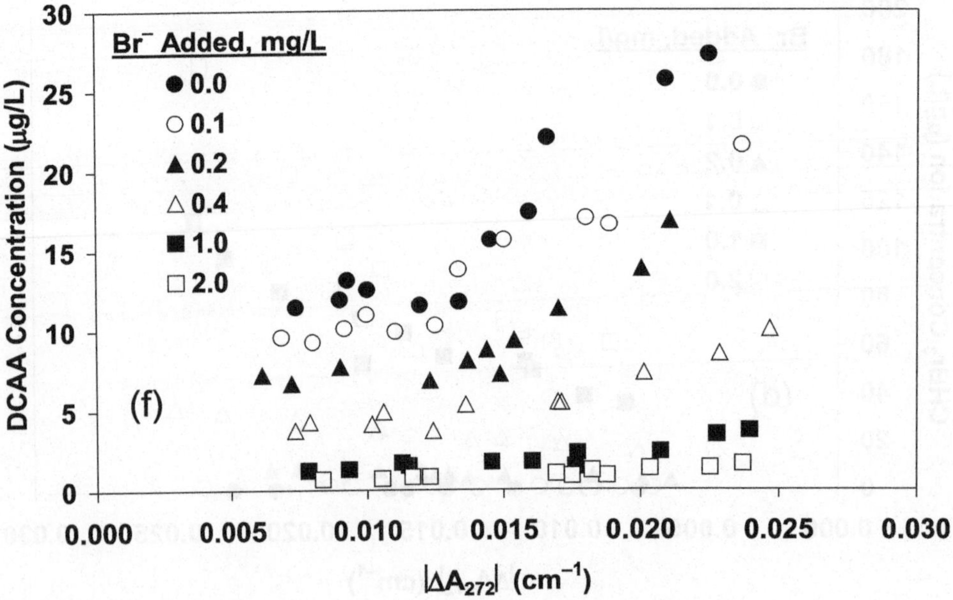

Figure 4.22f Relationship between DBP formation and $|\Delta A_{272}|$ in systems with varying Br^- concentration. LW water, Cl_2:DOC ratio 1.5, pH 7, 25°C. (f) DCAA.

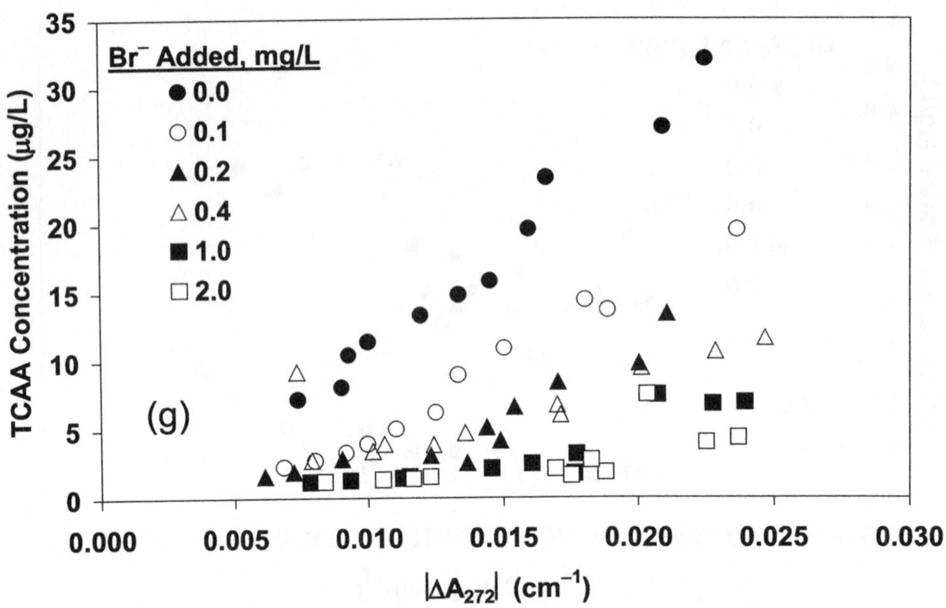

Figure 4.22g Relationship between DBP formation and $|\Delta A_{272}|$ in systems with varying Br^- concentration. LW water, Cl_2:DOC ratio 1.5, pH 7, 25°C. (g) TCAA.

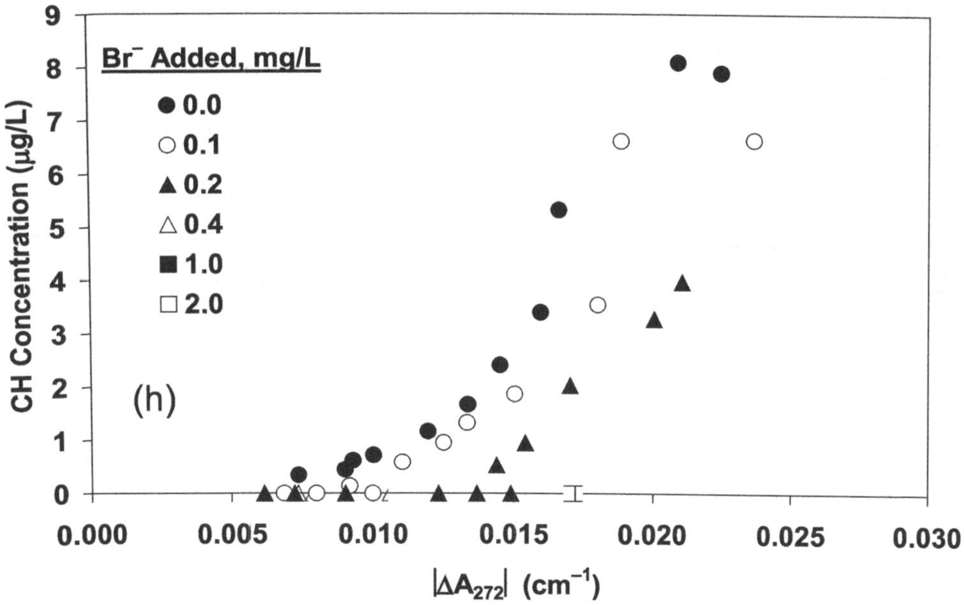

Figure 4.22h Relationship between DBP formation and $|\Delta A_{272}|$ in systems with varying Br^- concentration. LW water, Cl_2:DOC ratio 1.5, pH 7, 25°C. (h) CH.

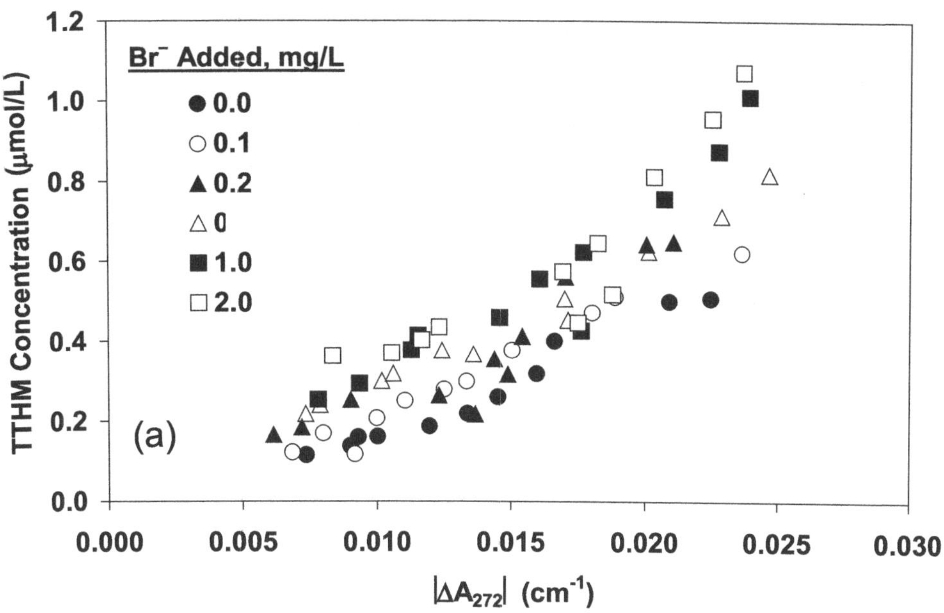

Figure 4.23a Relationship of TTHMs and THAAs with $|\Delta A_{272}|$ in LW water amended with varying Br^- concentrations. Cl_2:DOC ratio 1.5, pH 7, 25°C. (a) TTHMs.

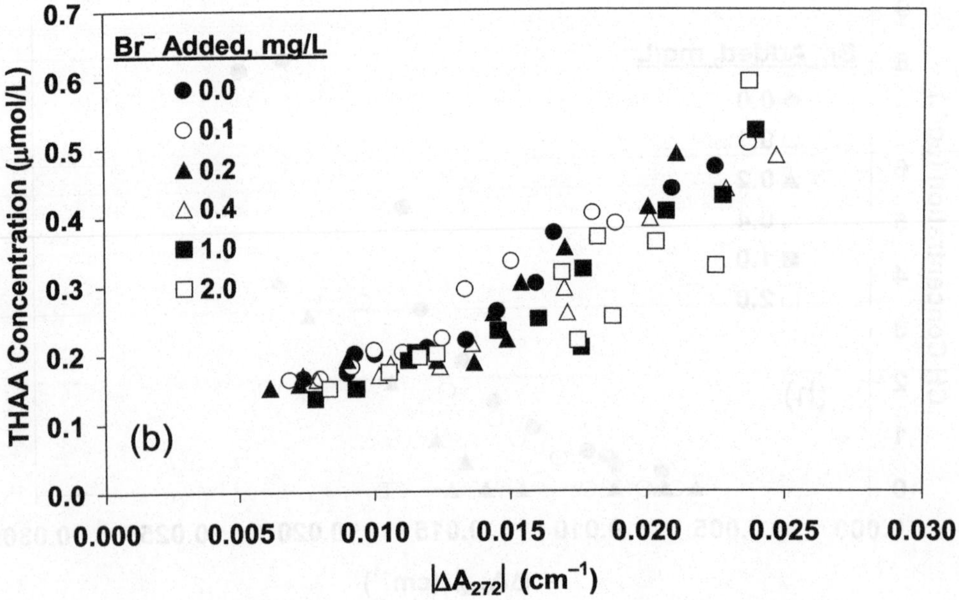

Figure 4.23b Relationship of TTHMs and THAAs with $|\Delta A_{272}|$ in LW water amended with varying Br^- concentrations. Cl_2:DOC ratio 1.5, pH 7, 25°C. (b) THAAs.

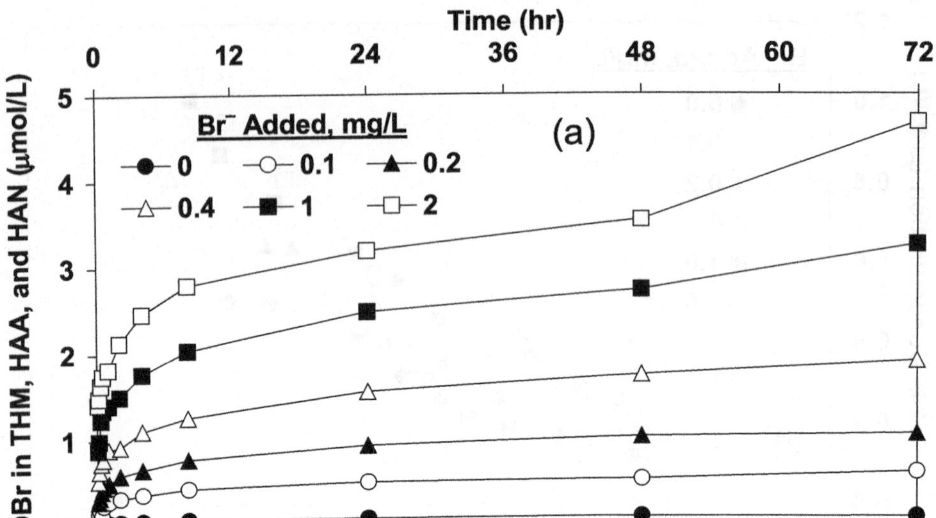

Figure 4.24a Total Br incorporated into THMs, HAAs, and HANs after chlorination of LW water with various Br^- additions. (a) Kinetics of formation of identified Br-containing DBPs. Cl_2:DOC ratio 1.5, pH 7, 25°C.

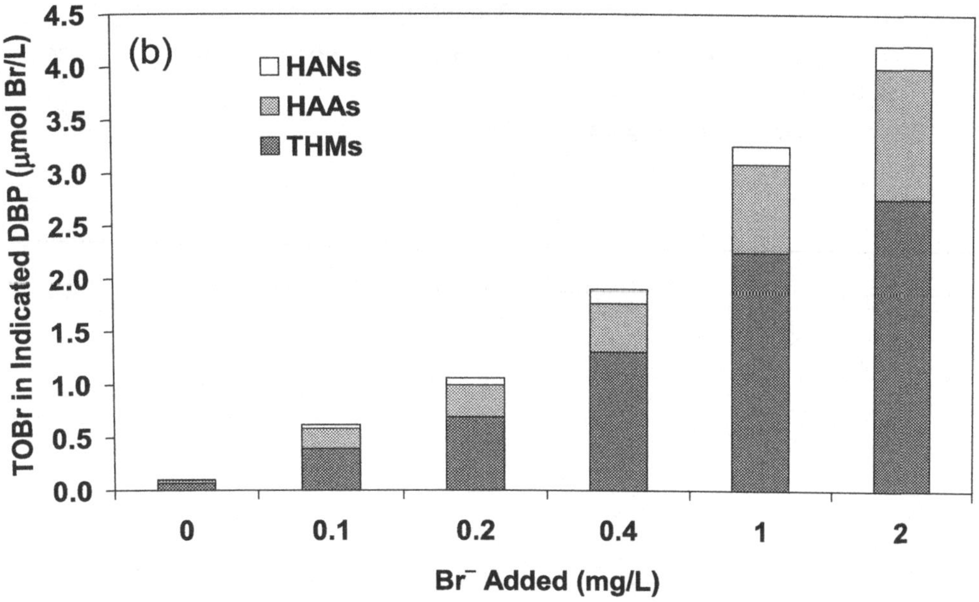

Figure 4.24b Total Br incorporated into THMs, HAAs, and HANs after chlorination of LW water with various Br^- additions. (b) Speciation of DBPs shown in part *a* at $t = 72$ hr. Cl_2:DOC ratio 1.5, pH 7, 25°C.

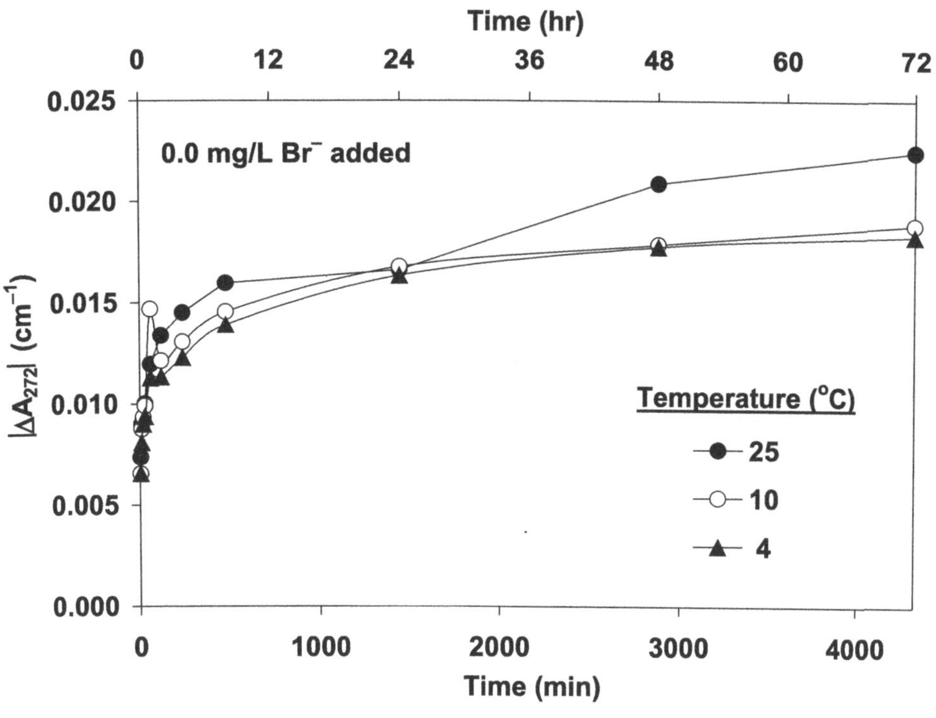

Figure 4.25a Effect of temperature on development of $|\Delta A_{272}|$ and chlorine consumption. LW water, Cl_2:DOC ratio 1.5, pH 7. (a) $|\Delta A_{272}|$.

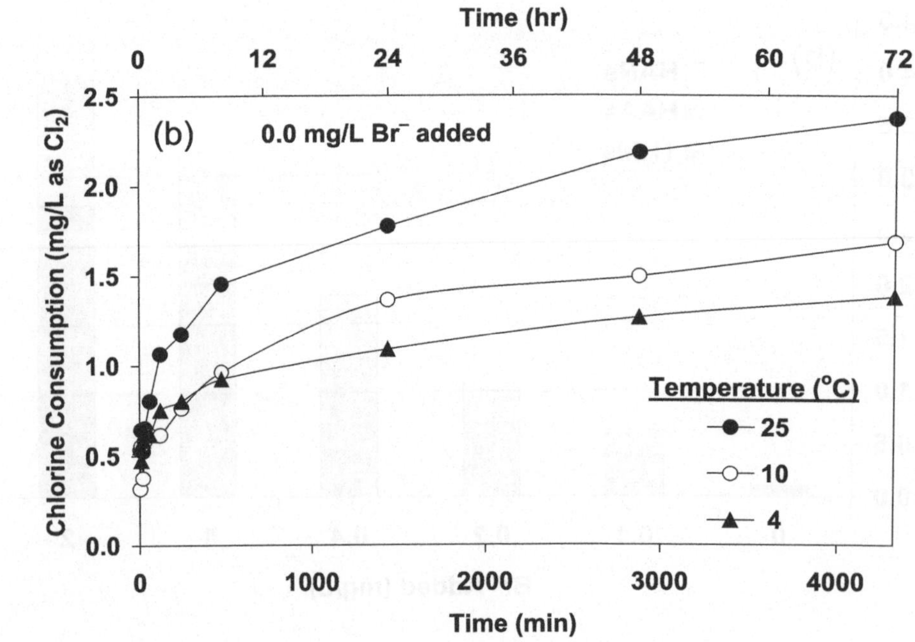

Figure 4.25b Effect of temperature on development of $|\Delta A_{272}|$ and chlorine consumption. LW water, Cl_2:DOC ratio 1.5, pH 7. (b) chlorine consumption.

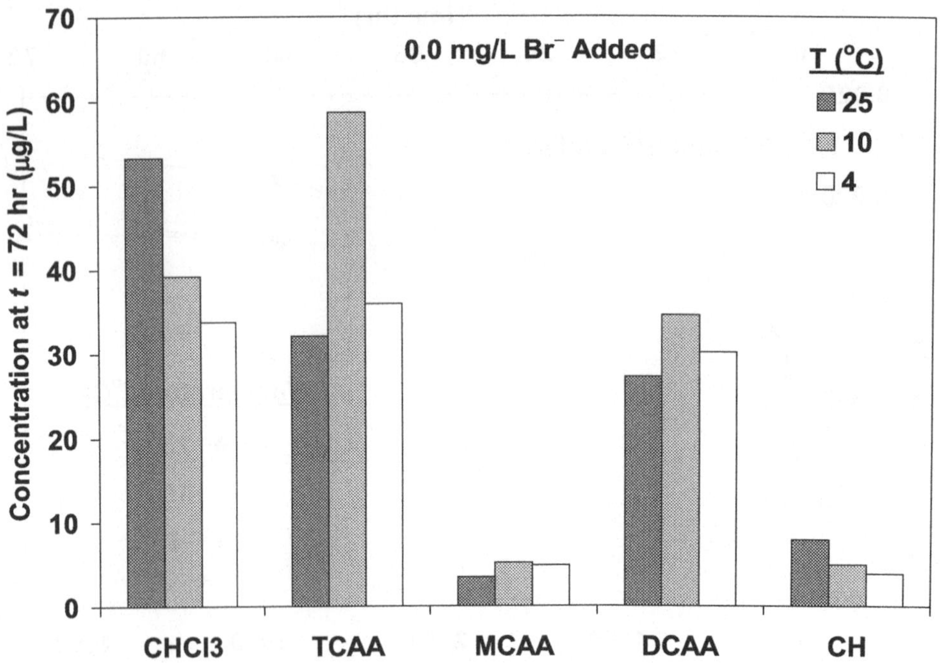

Figure 4.26a DBP formation at 25, 10, and 4°C, in systems with various additions of Br^-. LW water, Cl_2:DOC ratio 1.5, pH 7. (a) No Br^- added.

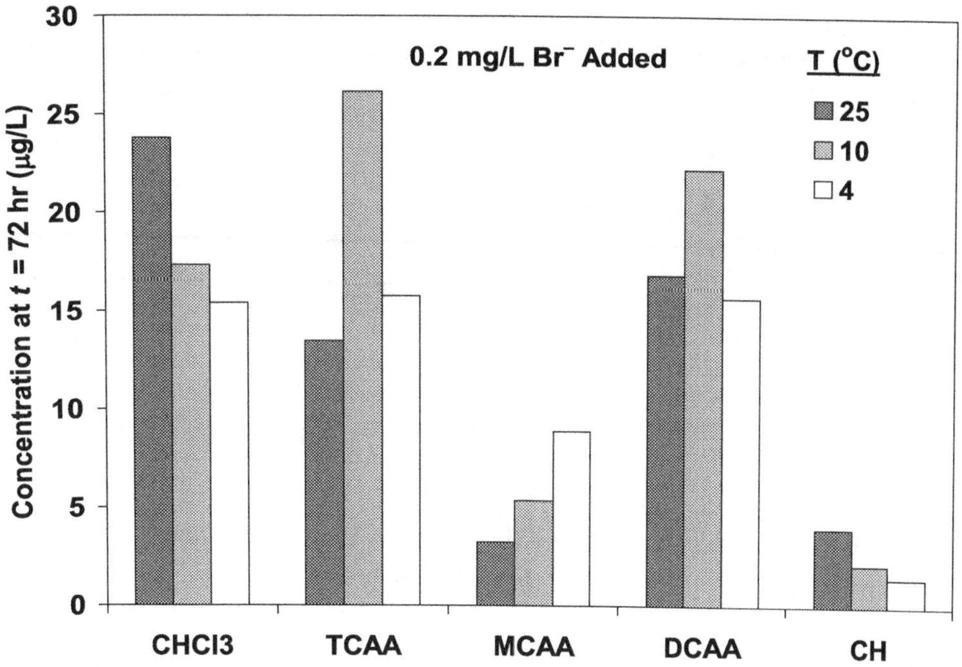

Figure 4.26b DBP formation at 25, 10, and 4°C, in systems with various additions of Br^-. LW water, Cl_2:DOC ratio 1.5, pH 7. (b) 0.2 mg/L Br^- added.

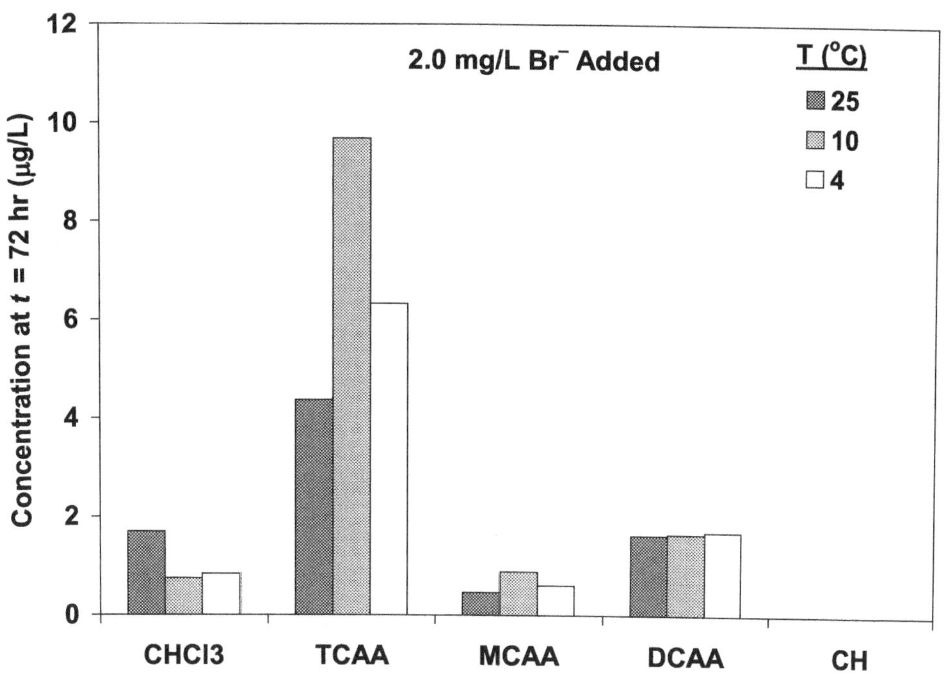

Figure 4.26c DBP formation at 25, 10, and 4°C, in systems with various additions of Br^-. LW water, Cl_2:DOC ratio 1.5, pH 7. (c) 2.0 mg/L Br^- added.

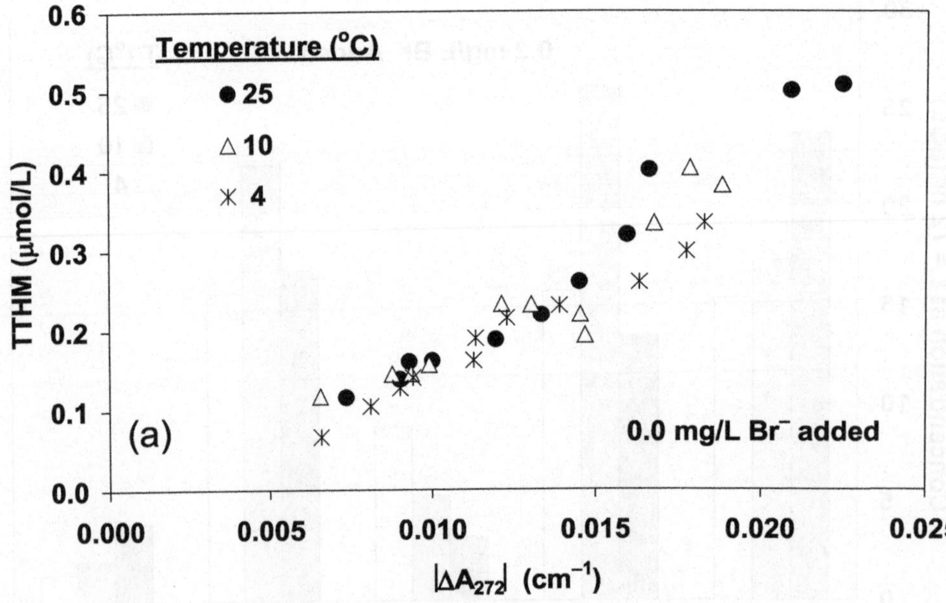

Figure 4.27a Relationship between TTHM formation and $|\Delta A_{272}|$ in LW water at various temperatures and Br^- concentrations. Cl_2:DOC ratio 1.5, pH 7. (a) No Br^- added.

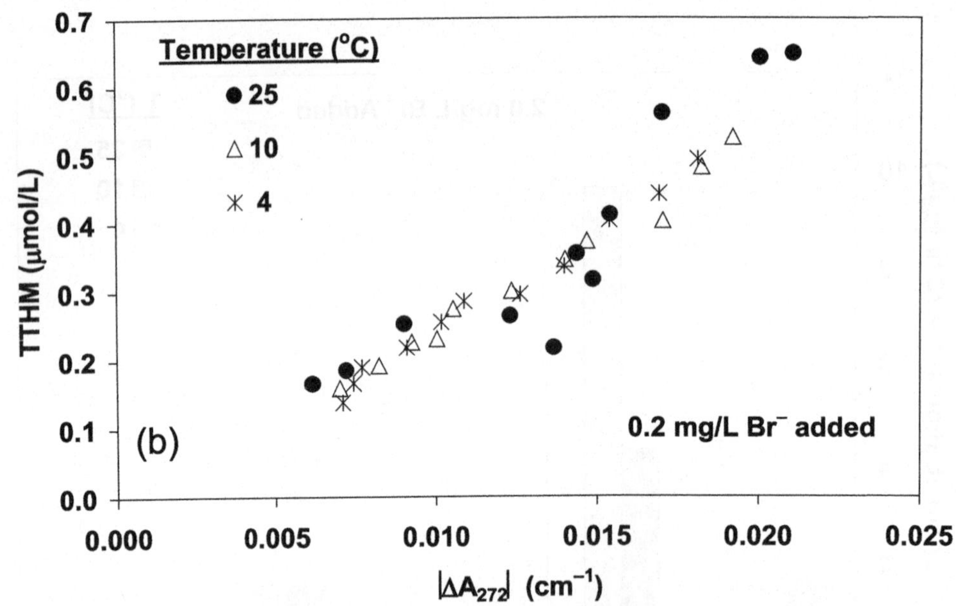

Figure 4.27b Relationship between TTHM formation and $|\Delta A_{272}|$ in LW water at various temperatures and Br^- concentrations. Cl_2:DOC ratio 1.5, pH 7. (b) 0.2 mg/L Br^- added.

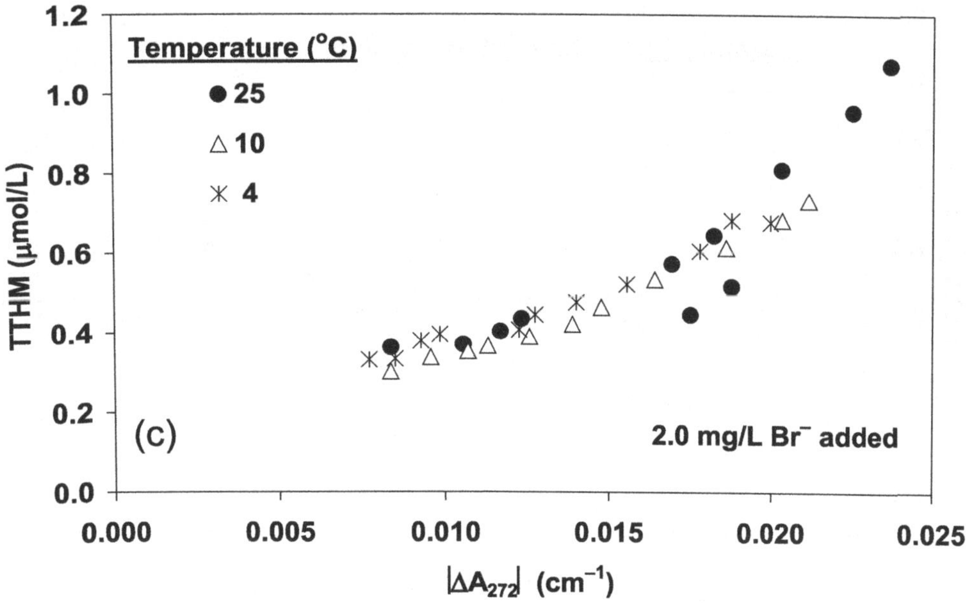

Figure 4.27c Relationship between TTHM formation and $|\Delta A_{272}|$ in LW water at various temperatures and Br^- concentrations. Cl_2:DOC ratio 1.5, pH 7. (c) 2.0 mg/L Br^- added.

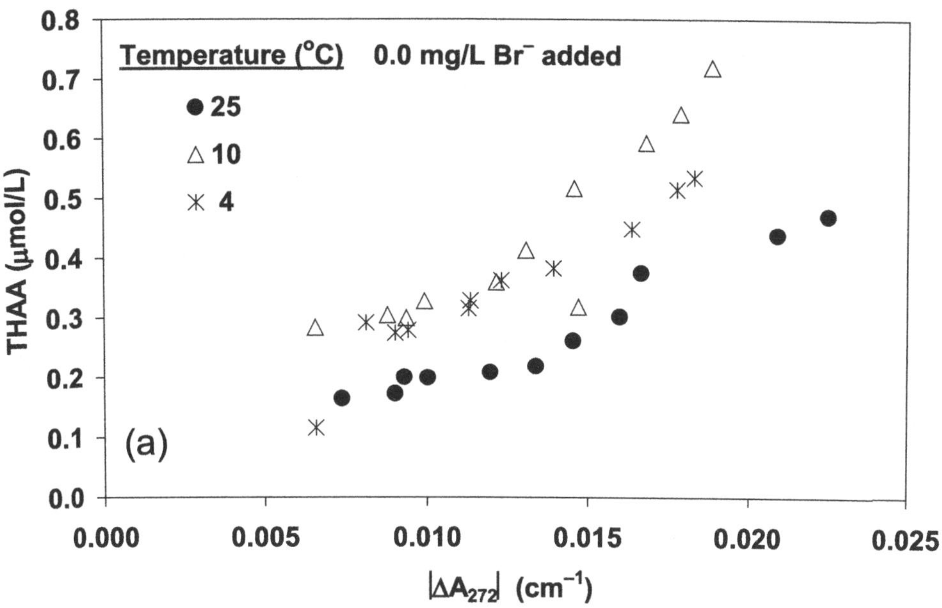

Figure 4.28a Relationship between THAA formation and $|\Delta A_{272}|$ in LW water at various temperatures and Br^- concentrations. Cl_2:DOC ratio 1.5, pH 7. (a) No Br^- added.

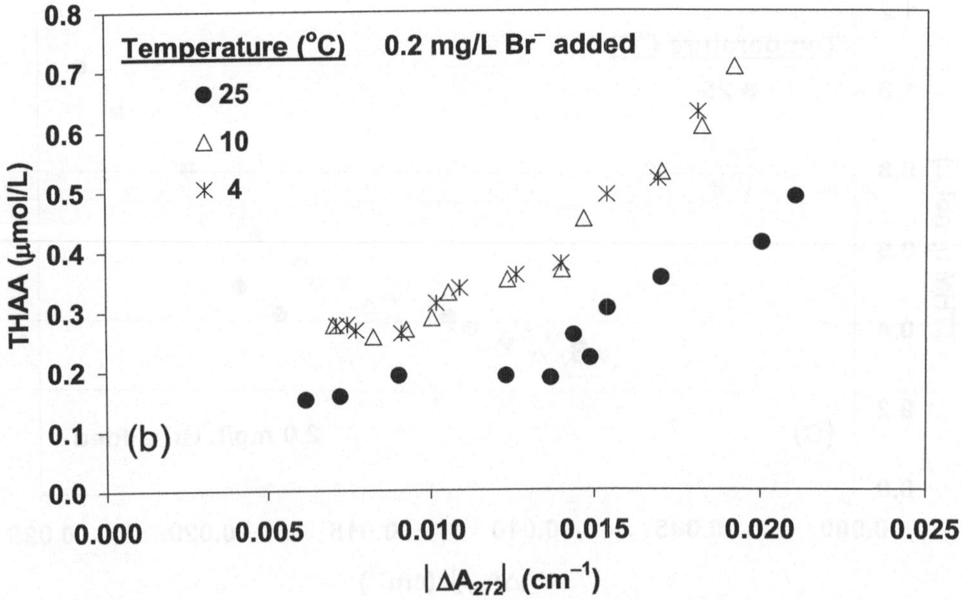

Figure 4.28b Relationship between THAA formation and $|\Delta A_{272}|$ in LW water at various temperatures and Br^- concentrations. Cl_2:DOC ratio 1.5, pH 7. (b) 0.2 mg/L Br^- added.

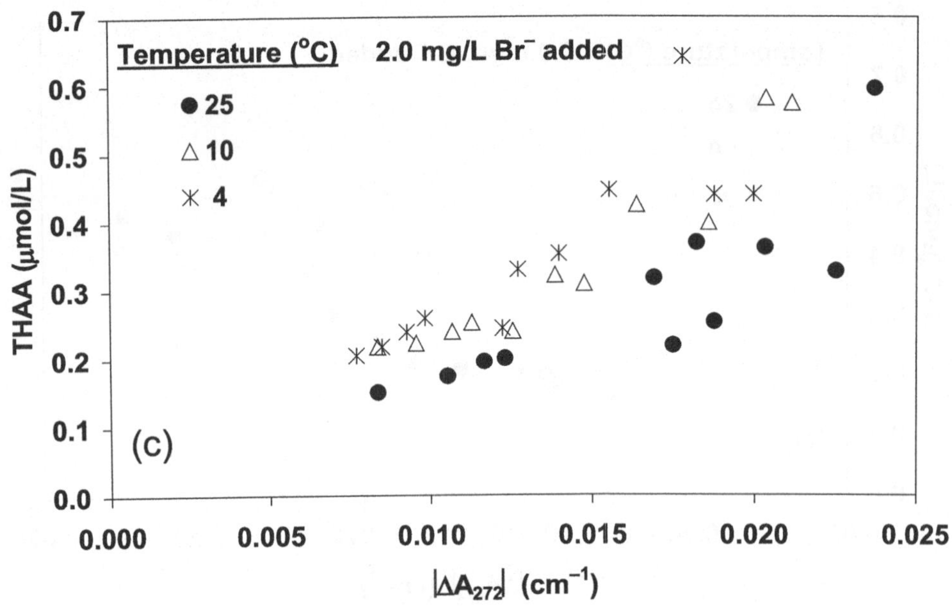

Figure 4.28c Relationship between THAA formation and $|\Delta A_{272}|$ in LW water at various temperatures and Br^- concentrations. Cl_2:DOC ratio 1.5, pH 7. (c) 2.0 mg/L Br^- added.

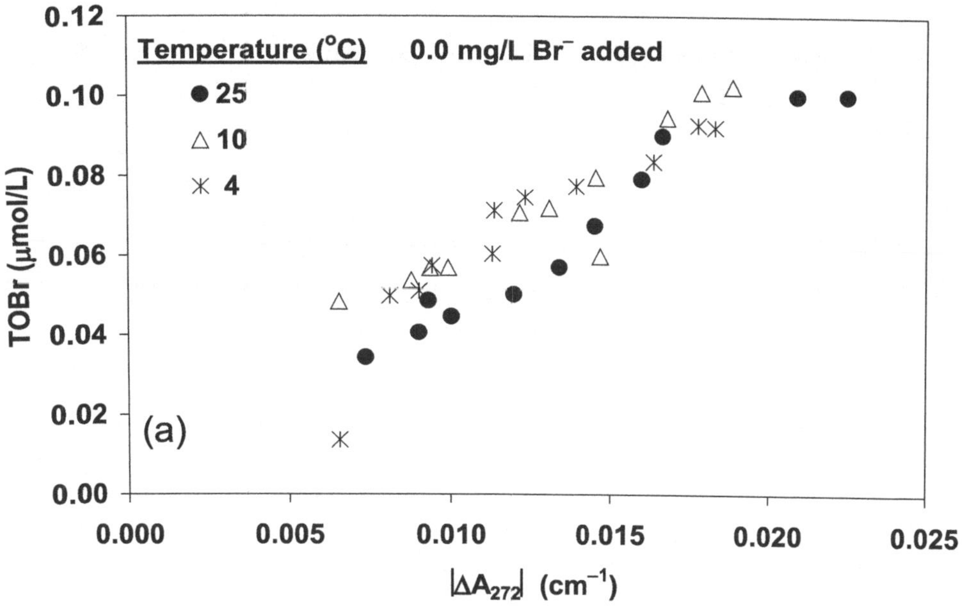

Figure 4.29a Relationship between Br incorporation into identified DBPs and $|\Delta A_{272}|$ in LW water at various temperatures and Br^- concentrations. Cl_2:DOC ratio 1.5, pH 7. (a) No Br^- added.

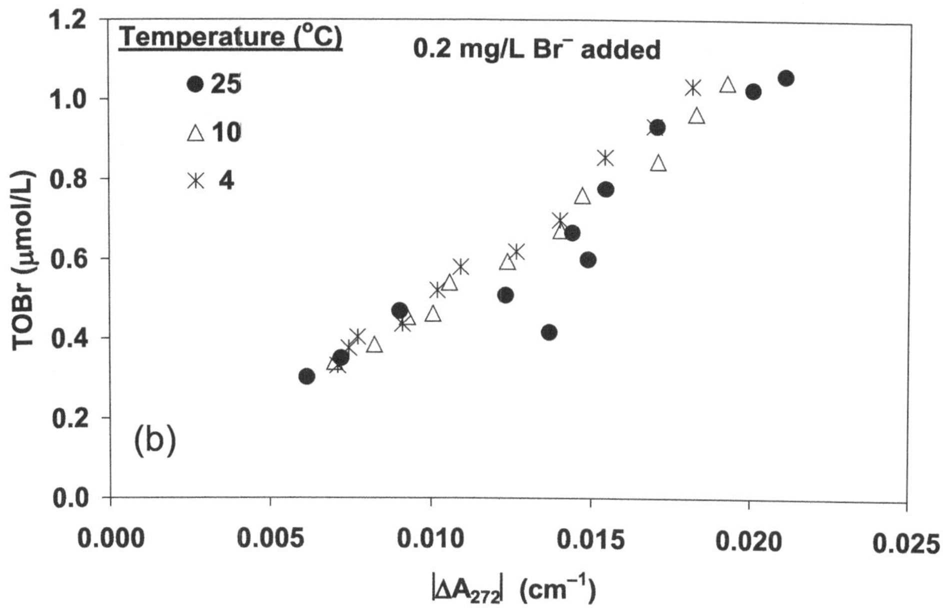

Figure 4.29b Relationship between Br incorporation into identified DBPs and $|\Delta A_{272}|$ in LW water at various temperatures and Br^- concentrations. Cl_2:DOC ratio 1.5, pH 7. (b) 0.2 mg/L Br^- added.

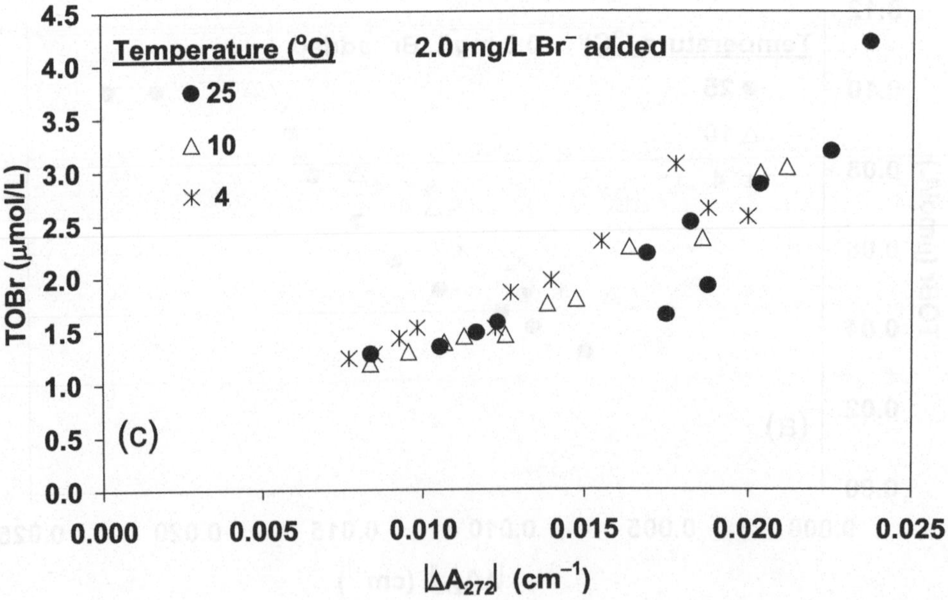

Figure 4.29c Relationship between Br incorporation into identified DBPs and $|\Delta A_{272}|$ in LW water at various temperatures and Br^- concentrations. Cl_2:DOC ratio 1.5, pH 7. (c) 2.0 mg/L Br^- added.

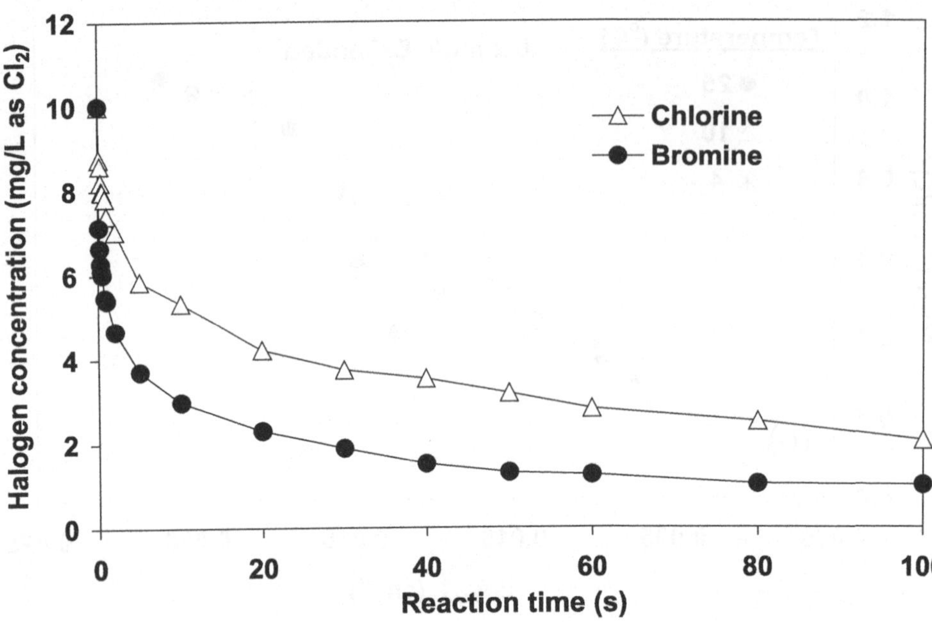

Figure 4.30 Oxidant decay in stopped-flow experiments for chlorination or bromination of concentrated NOM from LW. DOC 22.5 mg/L, pH 7.0. Oxidant dose, 10 mg/L as Cl_2 (1.41×10^{-4} M).

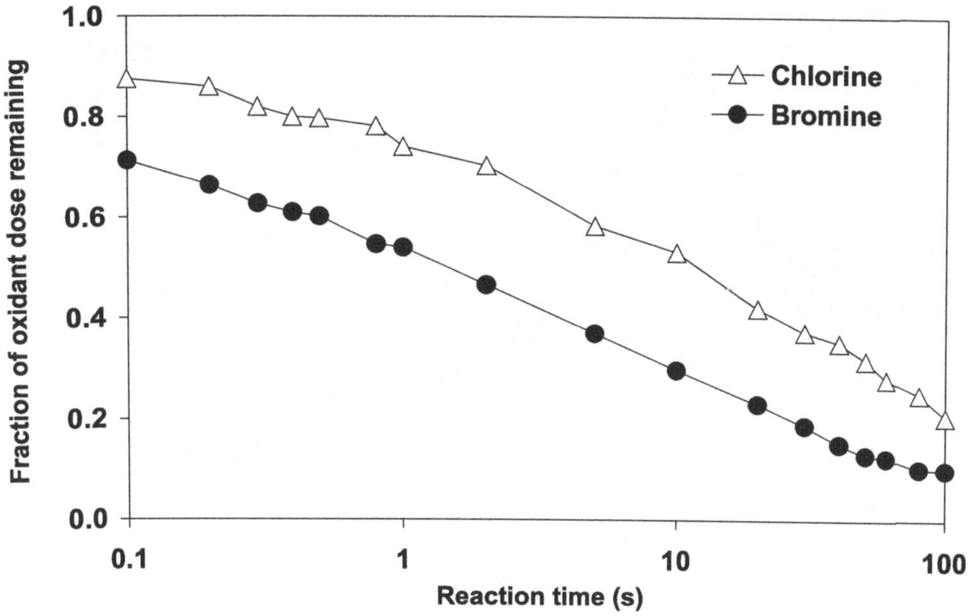

Figure 4.31 Data from Figure 4.30, replotted to show relative change in oxidant concentration and to show more clearly the data at very short reaction times

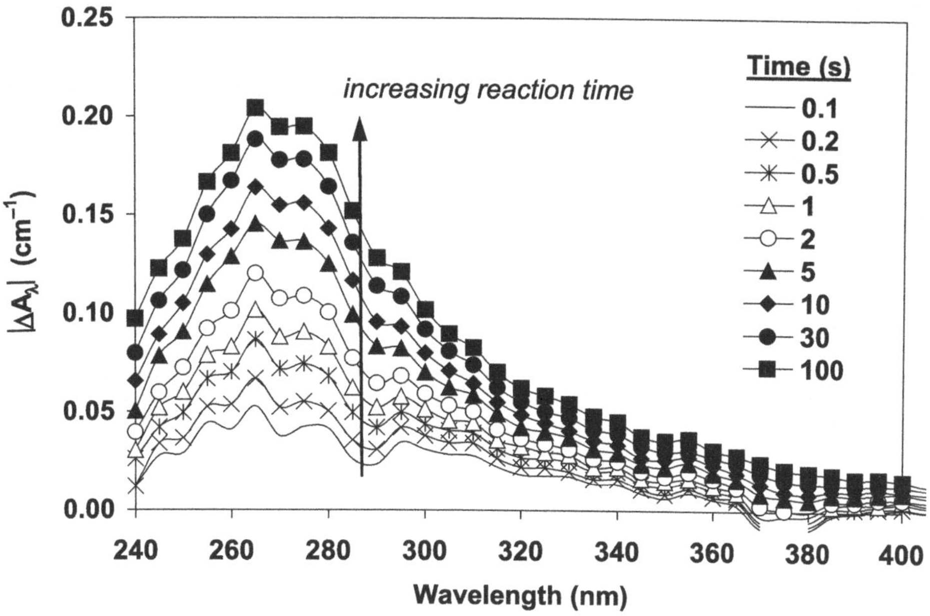

Figure 4.32 Differential spectra of chlorinated LW water after varying reaction times. Same conditions as in Figure 4.30.

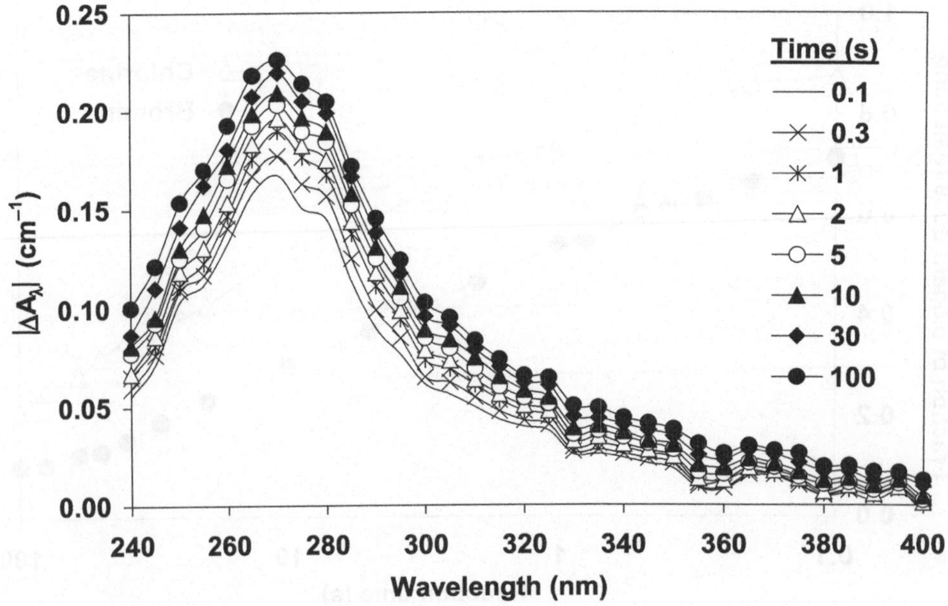

Figure 4.33 Differential spectra of brominated LW water after varying reaction times. Same conditions as in Figure 4.30.

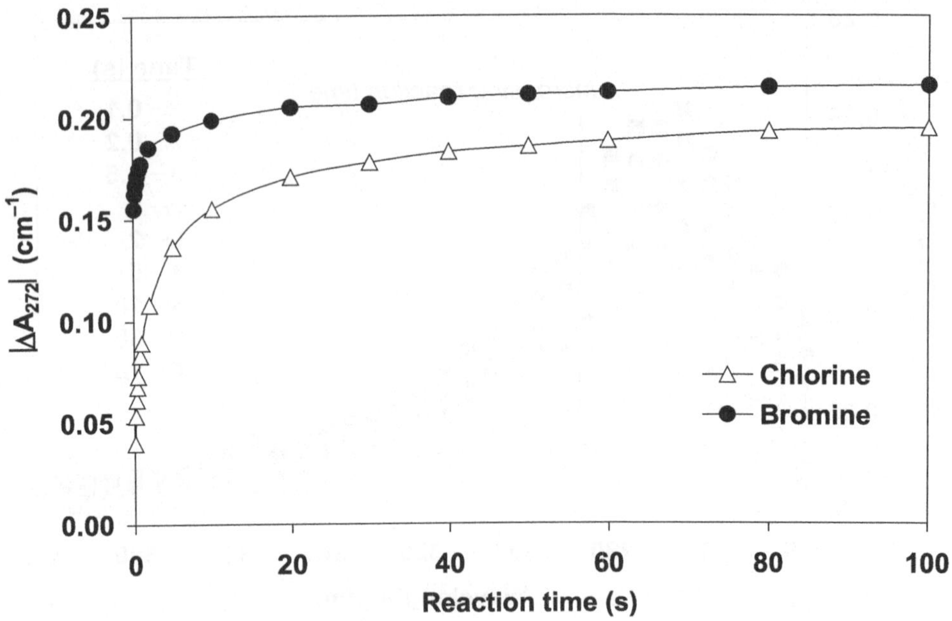

Figure 4.34 Development of differential absorbance over time in chlorinated or brominated LW water. Same conditions as in Figure 4.30.

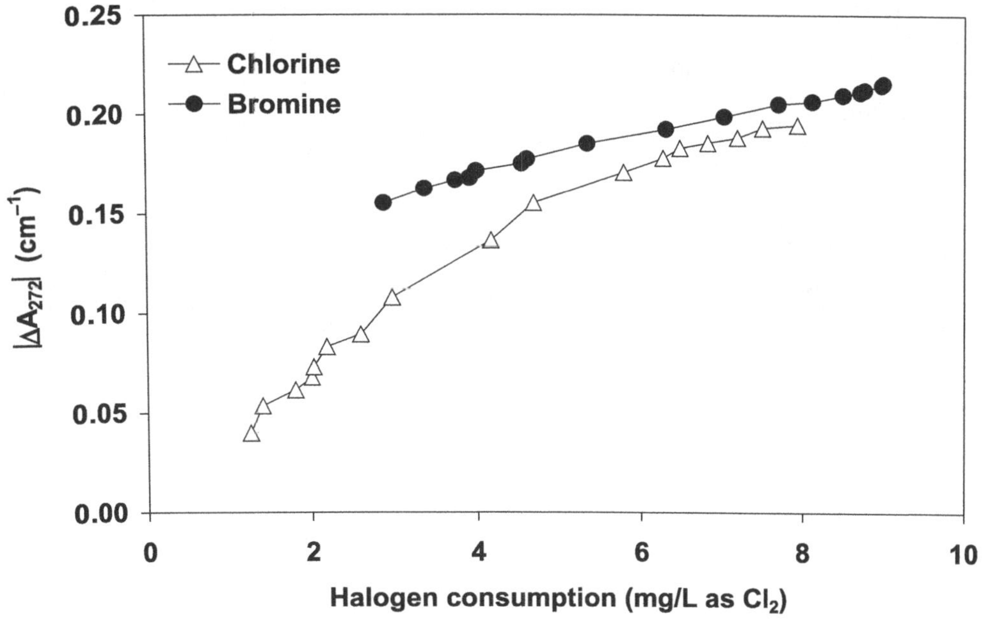

Figure 4.35 Relationship between differential absorbance and oxidant consumption in chlorinated or brominated LW water. Same conditions as in Figure 4.30.

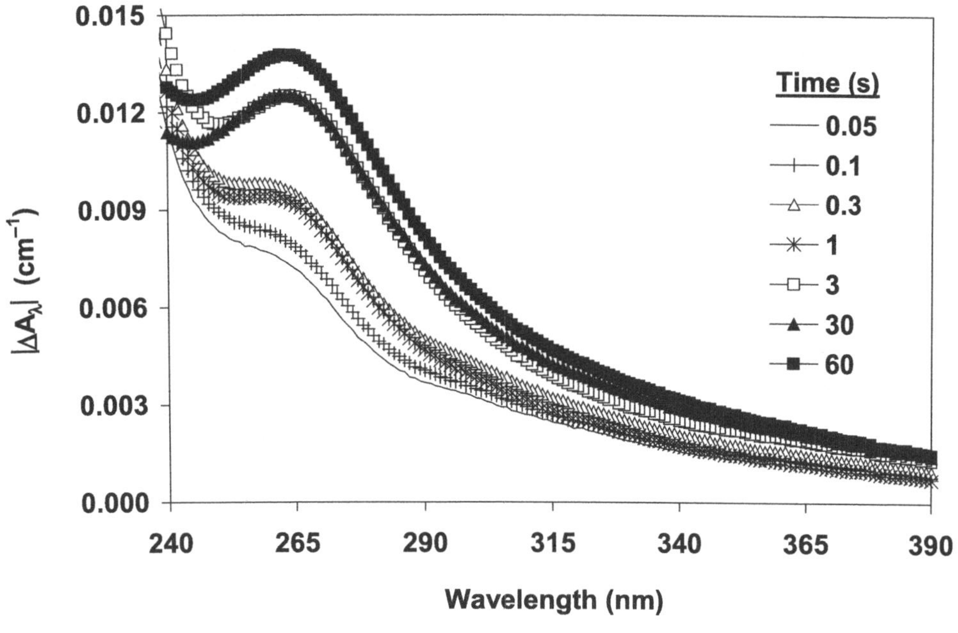

Figure 4.36 Differential spectra of LW NOM chlorinated in the quench-flow mode. 25 mg/L DOC, Cl_2:DOC ratio 1.5. All samples diluted to 2.2 mg/L DOC prior to analysis.

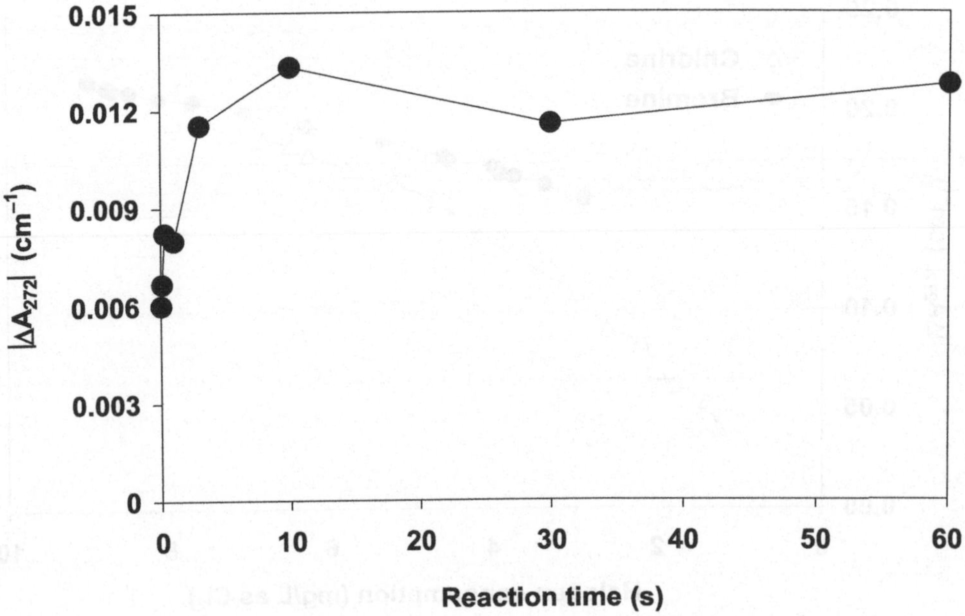

Figure 4.37 Kinetics of development of $|\Delta A_{272}|$ for chlorination of LW water at short reaction times in the quench-flow experiments. Same conditions as in Figure 4.36.

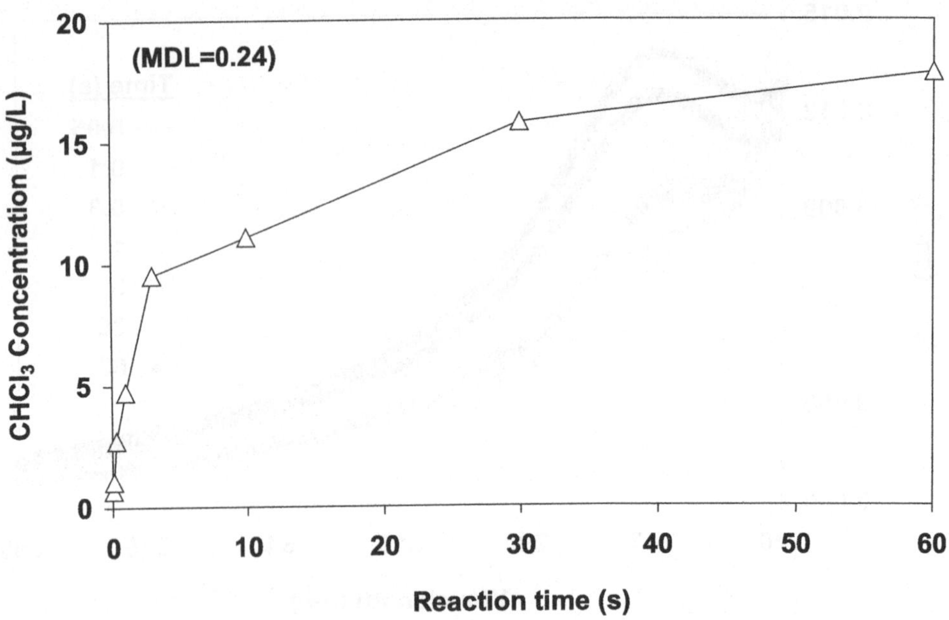

Figure 4.38 Kinetics of formation of $CHCl_3$ at short reaction times; other THMs were present at concentrations too far below the MDL to be significant. Same conditions as in Figure 4.36.

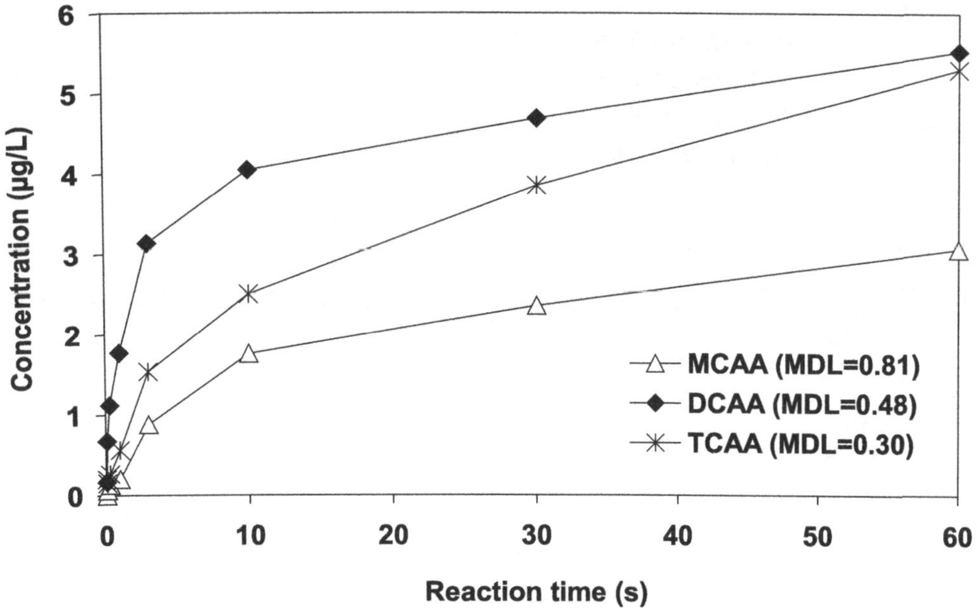

Figure 4.39 Kinetics of formation of HAAs at short reaction times. Same conditions as in Figure 4.36.

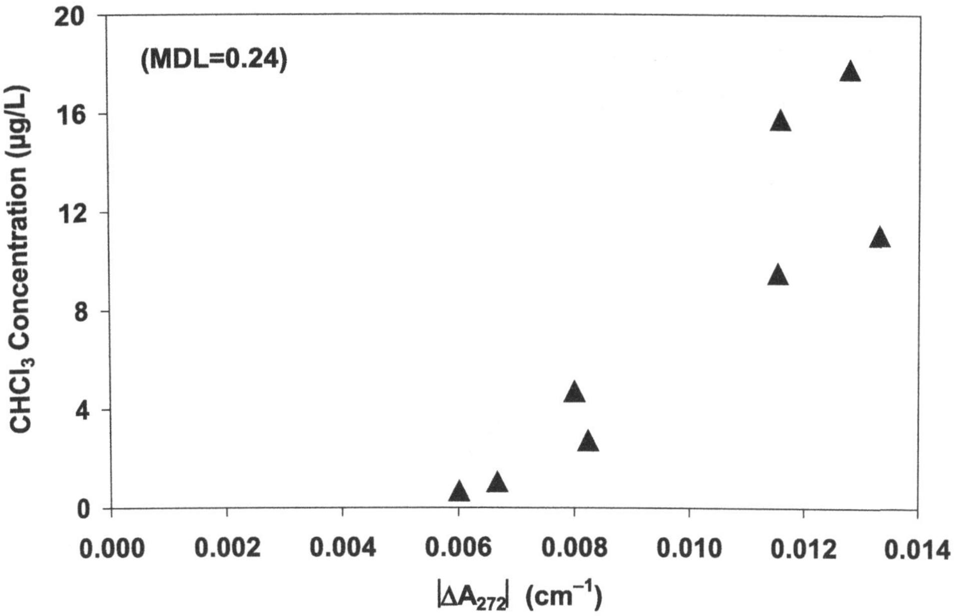

Figure 4.40 Relationship between $CHCl_3$ formation and differential absorbance for LW NOM chlorinated in the quench-flow mode; other THMs were present at concentrations too far below the MDL to be significant. Same conditions as in Figure 4.36.

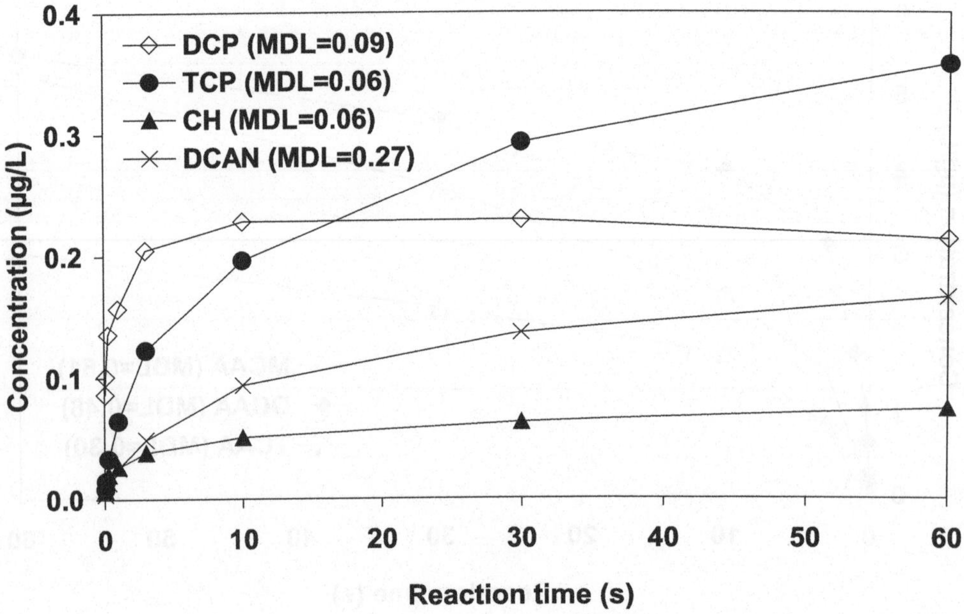

Figure 4.41 Kinetics of formation of several DBPs at short reaction times. Same conditions as in Figure 4.36.

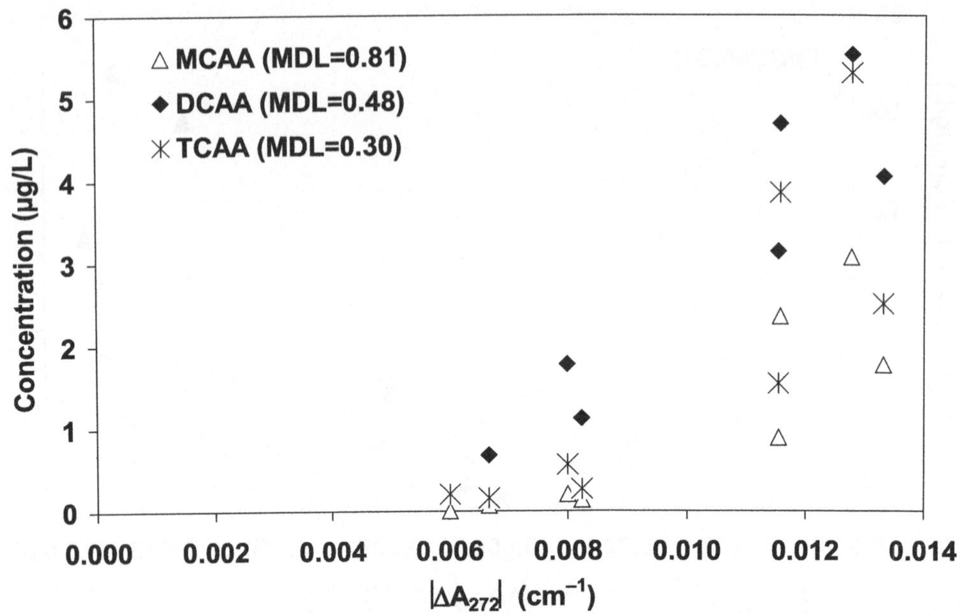

Figure 4.42 Relationship between HAA formation and differential absorbance in LW NOM chlorinated in the quench-flow mode. Same conditions as in Figure 4.36.

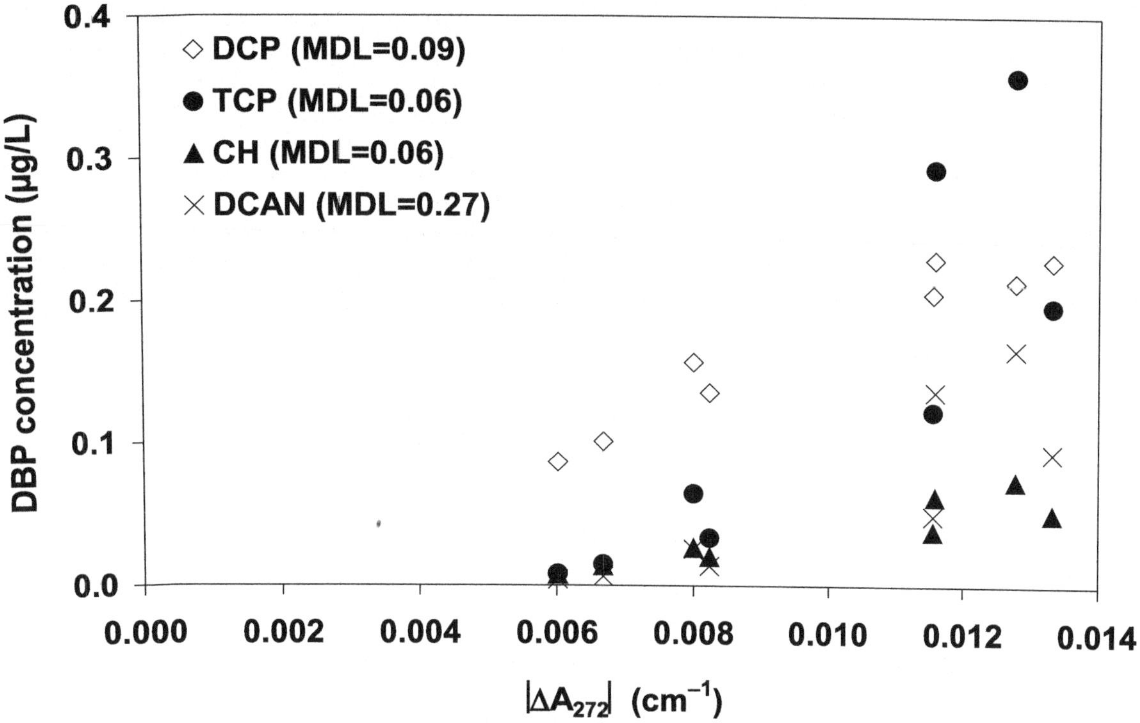

Figure 4.43 Relationship between formation of various minor DBPs and differential absorbance in LW NOM chlorinated in the quench-flow mode. Same conditions as in Figure 4.36.

CHAPTER 5
MODELING OF DBP FORMATION KINETICS

INTRODUCTION

As noted in the literature review, both statistical and mechanistic approaches have been used to model DBP formation. Statistical models correlate the experimental data with a range of independent parameters, without constraining the model equations to fit any particular theoretical framework. The number of fitting parameters is often large, providing enough degrees of freedom to fit most data sets. Mechanistic models, on the other hand, attempt to identify a set of reaction pathways and rate constants that fit the experimental data. Although this approach can provide insights into the chemistry of DBP formation, it has not been highly successful or widely adopted because of an insufficiency of data, particularly data for the concentration of precursor molecules or reaction sites in NOM.

In this chapter, an approach is described for developing a more complete and realistic mechanistic kinetic model for DBP formation during chlorination of NOM. The model was calibrated based on experiments with Lake Washington water and was then tested using several waters from across the U.S.

MODELING APPROACH

The assumed reaction scheme for the waters containing negligible Br^- concentrations is shown in Figure 5.1. In the figure, S, ICl_x, and PCl_3 represent, respectively, unreacted precursor sites, chlorinated intermediates in the reaction pathway, and a chlorinated product that does not release THMs or HAAs. (The species PCl_3 might react further to release other DBPs, such as chloral hydrate, but those DBPs are not included in this model.) The reactions designated by thick, solid arrows are chlorine incorporation reactions, and the reactions designated by arrows with broken lines do not incorporate more halogen into a molecule, but convert it to a non-reactive product (THM, HAA, or PCl_3).

All of the reactions in the scheme are assumed to be first-order with respect to the organic reactant (S or ICl_x) and $Cl_{2,tot}$ (total free chlorine), i.e., the rate expressions for all the reactions are of the following form:

$$r = k[Cl_{2,\,tot}][S \text{ or } ICL_x] \tag{5.1}$$

where, for each species i, the symbol $[i]$ represents the concentration of i in solution.

To obtain a reasonable fit between the model and experimental data, it was necessary to posit that NOM molecules contain at least two types of reactive sites, one of which (designated here as S_f) reacts more rapidly with chlorine than the other (S_s). The two types of sites were assumed to share the same reaction pathway, and to proceed along that path at different rates. In addition, it was assumed that the ratio of the rate constants for reaction of free chlorine with the two different types of sites was the same for all reaction steps, i.e., that

$$\frac{k_{f,\,1}}{k_{s,\,1}} = \frac{k_{f,\,2}}{k_{s,\,2}} = \frac{k_{f,\,3}}{k_{s,\,3}} = \ldots$$

where the subscripts 1, 2, 3, etc., represent different steps in the reaction pathway.

The model calibration effort consisted of identifying the parameter values that generated the best fit between the predicted and measured values of all THMs and HAAs. In theory, it would have been possible to model the kinetics of overall chlorine consumption as well. However, to do so, a separate branch of the reaction scheme would have been required to model chlorine consumption via reactions that did not generate chlorinated DBPs. Rather than attempting to model those reactions, the free chlorine concentration during each experiment was fit to an equation of the form $[Cl_{2,tot}] = b_1 \ln t + b_2$, where b_1 and b_2 are empirical constants. The fit of the model equation to the chlorine decay data was very good, as shown for several experiments in Figure 5.2. The model equation describing the $Cl_{2,tot}$ concentration over time was then used as an input parameter in the DBP kinetics model, so that the simulated chlorine concentration always followed that determined experimentally.

The DBP model was calibrated using data from chlorination of Lake Washington water. Because the initial Br^- concentration in that water could not be quantified, only Br-free DBPs were considered when these results were modeled. The values of all eight rate constants shown in Figure 5.1 were allowed to fluctuate, along with the total concentration of reactive sites at time zero ($[S_{tot}] = [S_S]_0 + [S_f]_0$), the fraction of total sites that were slow-reacting (i.e., $[S_S]_0/[S_{tot}]_0$), and the ratio of the rate constants of the fast- and slow-reacting sites (k_f / k_s). The fitting exercise was carried out using the *fminsearch* option in MATLAB®.

A similar approach was used to model systems in which bromination was significant (chlorination of LW water with bromide added or of other source waters containing Br^- at significant levels). The reaction pathways that were considered in these systems are shown in Figure 5.3.

In these systems, all of the Br^- in the solution was assumed to be converted to HOBr immediately upon addition of the free chlorine. All halogen incorporation reactions were assumed to be first-order with respect to the organic reactant (S or ICl_xBr_y) and the free halogen (either $Cl_{2,tot}$ or $Br_{2,tot}$), so the rate expressions for those reactions were as follows:

For chlorine incorporation: $$r_{ICl_{x+1}Br_y} = k_{ICl_{x+1}}[Cl_{2,\,tot}][\text{S or } ICl_xBr_y] \quad (5.2)$$

For bromine incorporation: $$r_{ICl_xBr_{y+1}} = k_{ICl_{x+1}}[Br_{2,\,tot}][\text{S or } ICl_xBr_y] \quad (5.3)$$

The reactions forming end products were also assumed to be first-order with respect to each of the reactants (the organic intermediate and the oxidant), but these reactions were presumed to be mediated by either oxidant in the system. Therefore, the rates of these reactions could be described by the following composite rate expression that combines the effects of the two reactions in parallel (one with $Cl_{2,tot}$ and one with $Br_{2,tot}$):

For forming end products: $$r_{\substack{\text{end}\\\text{product}}} = k_{\substack{\text{end}\\\text{product}}}([Cl_{2,\,tot}] + a[Br_{2,\,tot}])[ICl_xBr_y] \quad (5.4)$$

In the above expression, a is the ratio of the rate constants for the release of the DBP by oxidation of the halogenated intermediate with $Br_{2,tot}$ and $Cl_{2,tot}$, respectively.

In the systems containing significant Br^- concentrations, the number of adjustable parameters was larger than could reasonably be estimated, given the available data. Therefore, assumptions were made to reduce that number. First, the relative rates of bromination and chlorination were assumed to be the same for all reactions in the reaction scheme, except for the reactions forming the final products. (The corresponding ratio for formation of end products was designated a, as note above.) The ratio of these rates, k_{Br} / k_{Cl}, is referred to as the branching ratio and is designated subsequently as R. Second, the rate constants for chlorination of all species with a given degree of halogenation were assumed to be identical, as were the rate constants for bromination. Thus, for instance, the rate constant for chlorination of ICl_xBr_y was assumed to be the same for all dihalogenated species (i.e., those with $x + y = 2$: ICl_2, $IClBr$, and IBr_2). These assumptions allowed the systems in which bromination occurred to be modeled by using the 11 rate constants for a system with chlorination only, the relative rates of formation of end products via oxidation with bromine versus chlorine (the parameter a in Equation 5.4), and the branching ratio, R (i.e., a total of 13 adjustable parameters). Note that, for modeling the data for chlorination of LW water with Br^- added, 11 of these parameters were pre-determined, based on the data from systems containing chlorine as the only halogen).

MODEL RESULTS AND INTERPRETATION

Waters With Very Low Br^- Concentrations

The optimized parameter values from the calibration exercise (LW water, no Br^- added), are shown in Table 5.1, and the fit of the model to the data for all the DBPs at a Cl_2:DOC dosing ratio of 1.5 mg/mg is shown in Figure 5.4. The model predictions and experimental data for $CHCl_3$ formation in the waters at all other dosing ratios investigated are shown in Figure 5.5. The model reproduces the experimental results quite well at the dosing ratio of 1.5 throughout the time frame of the experiments for $CHCl_3$, DCAA, and TCAA; the fit to the experimental data for MCAA at this dosing ratio was not as good. The model also fits the data for $CHCl_3$ at the other dosing ratios fairly well, except the lowest dosing ratio investigated (0.1), in which case the observed $CHCl_3$ concentrations were substantially greater than the predicted concentrations (although both sets of values were extremely low).

The relative contributions of the fast- and slow-reacting sites to the formation of the various DBPs, according to the simulation, are included in the figures. The simulations suggest that, at Cl_2:DOC ratios less than ~0.4, all the chlorine is consumed rapidly, so that the DBPs arise primarily from the rapidly reacting sites. At higher ratios, the chlorine persists long enough for the slowly reacting sites to contribute substantially to the ultimate DBP formation.

The rate constant for the first chlorination step was orders of magnitude less than that of any other step, and, since all the rate expressions have the same form, this result indicates that the initial chlorination step is rate-limiting. The best-fit value for the total concentration of reactive sites in these experiments was 3.0 μmol/L, corresponding to approximately 0.012 mole of sites per mole of DOC. Approximately 28% of these sites are "rapid reaction" sites, which, according to the model, have rate constants that are 340 times those of the slowly reacting sites.

For comparison, Gallard and von Gunten (2002) estimated that 17% to 28% of the sites in NOM were fast-reacting sites in their model of THM formation in various source waters, and Gang et al. [2002, 2003] estimated the ratio of rate constants for fast- versus slow-reacting sites to be between 90 and 145, based on their modeling efforts.

As a very rough check of the reasonableness of the value of $[S_{tot}]_0$, one can assume that each aromatic ring contains one reactive site. In such a case, since each ring contains six carbon

Table 5.1
Optimized reaction rate constants for chlorination of Lake Washington water at a dosing ratio of Cl_2:DOC = 1.5

Parameter	Value
Rate Constants for Slow Sites*	
k_{ICl}	1.5×10^{-5}
k_{ICl2}	3.3
k_{ICl3}	1.0×10^{-2}
k_{PCl3}	2.0
k_{MCAA}	0.10
k_{DCAA}	1.5×10^{-3}
k_{TCAA}	0.62
k_{CHCl3}	0.74
Other Fitting Parameters	
$S_{s,0} / S_{tot,0}$	0.72
k_f / k_s	340
Total Reactive Sites†	3.0

*units for all rate constants are $min^{-1}(\mu mol/L)^{-1}$
†in μmol/L

atoms, and the model suggests that the NOM has 0.012 mole of sites per mole of DOC, the NOM would contain 6×0.012, or 0.072 mole of aromatic carbon per mole of DOC, i.e., 7.2% of the carbon atoms in NOM would be in aromatic rings. Estimates of this value from independent measurements suggest that it is in the range 5–20% (e.g., Traina, Novak, and Smeck 1990; Reckhow, Singer, and Malcolm 1990; Croué, Korshin, and Benjamin 2000), so the estimate obtained from the fitting exercise is at least reasonable.

As noted above, the simulations at the lowest Cl_2:DOC ratio (0.1 mg/mg) significantly underestimated the observed $CHCl_3$ formation (Figure 5.5a). Although the data are not shown, the same statement applies to most other DBPs at this dosing ratio. The easiest way to reduce this deviation is to assume that a third, very rapidly reacting type of site is present at a very low concentration. Such a site could account for significant DBP formation at the very early stages of reaction, but, because it is present at a very low concentration, the error associated with ignoring it would be small in cases where the Cl_2:DOC ratio is > 0.2. Therefore, its existence (if, indeed, the explanation is correct) is not of much practical importance for modeling DBP formation in most drinking water treatment systems. Nevertheless, it is worth noting that inclusion of this group of sites is apparently necessary to model DBP formation over the entire range of conditions investigated.

The performance of the model when applied to LW water dosed at a Cl_2:DOC ratio of 1.5 at different pH values mirrored the results at pH 7 very closely. At pH values of 6, 8, and 9, just as at pH 7, some MCAA and DCAA formed very rapidly, but the subsequent generation of these species was very slight. The model was unable to reproduce such a rapid initial formation followed by very slow subsequent generation of these species. The model results are not shown, because they were essentially identical to those shown in Figures 5.4b and 5.4c.

The general patterns of $CHCl_3$ and TCAA generation were also the same at pH 6, 8, and 9 as at pH 7. However, as noted previously, $CHCl_3$ formation increased significantly with increasing

Table 5.2
Optimized rate constants for final DBP formation reaction at various pH values, for Lake Washington water at Cl_2:DOC = 1.5

Rate constant	pH 6	pH 7	pH 8	pH 9
k_{MCAA}	0.12*	0.10	0.13	0.13
k_{DCAA}	1.6×10^{-3}	1.5×10^{-3}	1.6×10^{-3}	1.7×10^{-3}
k_{TCAA}	0.60	0.62	0.41	0.26
k_{CHCl3}	0.57	0.74	0.82	1.02
k_{PCl3}	2.46	2.05	1.88	2.17

*units for all rate constants are $min^{-1}(\mu mol/L)^{-1}$

pH, and TCAA formation decreased, but less dramatically. These changes could be reproduced by allowing the rate constant for the final step in the respective reaction sequences to vary with pH. The model results for overall formation of these species are compared with the experimental data in Figure 5.6. Although the contributions of the fast- and slow-reacting sites to the overall formation of these species were calculated, they are not shown, to reduce the clutter in the figure. The relative contributions of the two types of sites were very similar to those at pH 7, shown in Figures 5.4a and 5.4d. The rate constants used in these simulations are listed in Table 5.2.

When the model was applied to other waters with Br^- concentrations below the MDL, the initial concentrations of reactive sites in those waters were estimated by assuming that the ratios $[S_{tot}]_0/A_{254}$ and $[S_{tot}]_s/[S_{tot}]_0$ were the same as in Lake Washington water, and that the rate constants for all the reactions were identical to those for Lake Washington water. (An assumption that $[S_{tot}]_0$ was proportional to A_{272} instead of A_{254} would have led to essentially the same result; the value of A_{254} was chosen based on its historical use as a measure of the aromatic content of NOM.) The simulations matched the general trends in the data, but the fit was only fair to poor. Therefore, optimal rate constants for the initial and final steps in each DBP-forming sequence (the step forming the DBP) were adjusted for each water, while keeping the constants for the other reactions (formation of intermediate species or PCl_3) the same.

The optimized rate constants for the various water sources with low Br^- are shown in Figure 5.7 and Table 5.3. Most of the reaction rate constants are within 0.5 log units (i.e., a factor of 3) of the values for Lake Washington; the only exceptions are k_{TCAA} and k_{CHCl3} for the sample with the lowest A_{254} value (Little Cottonwood Creek). Using the rate constants shown in the figure, the simulated and experimental data match very well, as shown for TCAA and $CHCl_3$ in Figure 5.8.

To summarize this portion of the modeling, a two-site reaction scheme was developed that successfully simulated formation of chloroform and all chlorinated HAAs from chlorination of several natural waters with low bromide concentrations, treating the concentration of reactive sites in the raw water as being proportional to A_{254}. The fit of the model to the data could be improved by assuming that a third, very rapidly reacting site is present at low concentrations and accounts for the majority of DBP formation at very low Cl_2:DOC ratios or very short reaction times. The optimized reaction rate constants suggest that the rate-limiting step in all cases is the first chlorine attachment step. The rate constants for different reactions in the sequence vary by approximately five orders of magnitude, but the rate constants for a given step vary by only approximately a factor of three among the different water sources tested.

Table 5.3
Optimized reaction rate constants for chlorination of various source waters with low Br^- concentrations (Cl_2:DOC = 1.5)

Parameter	Little Cottonwood	Eagle Mt. Lake	Florence WTP	Minneapolis	Narrows WTP	Sioux Falls	St. Paul WTP
Rate Constants for Slow Sites							
k_{ICl}	7.2×10^{-6}	6.3×10^{-6}	1.1×10^{-5}	1.9×10^{-5}	1.4×10^{-5}	7.9×10^{-6}	6.1×10^{-6}
k_{ICl2}*	3.3	3.3	3.3	3.3	3.3	3.3	3.3
k_{ICl3}*	1.0×10^{-2}	1.0×10^{-2}	1.0×10^{-2}	1.0×10^{-2}	1.0×10^{-2}	1.0×10^{-2}	1.0×10^{-2}
k_{PCl3}*	2.0	2.0	2.0	2.0	2.0	2.0	2.0
k_{MCAA}	0.23	0.24	1.5	0.56	1.2	1.0	1.1
k_{DCAA}	1.1×10^{-2}	3.0×10^{-3}	1.6×10^{-3}	8.0×10^{-4}	1.3×10^{-3}	1.3×10^{-3}	1.3×10^{-3}
k_{TCAA}	2.1×10^{5}	0.31	0.42	0.25	0.36	0.37	0.40
k_{CHCl3}	3.2×10^{5}	0.46	0.60	0.42	0.50	0.51	0.70
Other Fitting Parameters							
$S_{s,0}/S_{tot,0}$*	0.72	0.72	0.72	0.72	0.72	0.72	0.72
k_f/k_s*	340	340	340	340	340	340	340
Total Sites	0.53	4.6	6.4	17	6.5	10	8.8

*Held constant at the value determined in experiments with LW water

Waters With Significant Br^- Concentrations

As noted previously, for waters in which significant amounts of brominated DBPs form, the model was adjusted to include two additional fitting parameters: the branching ratio, *R*, and the ratio of the rates of oxidation of the various intermediates to final products, *a*. The data for all the DBPs that were detected after chlorination of LW water amended with 0.4 mg/L Br^- were used to calibrate the model with respect to these parameters, yielding best fit values of 12 and 11 for *R* and *a*, respectively. However, when these values were applied to model DBP formation in LW water amended with different concentrations of Br^-, the predicted DBP profiles did not conform well to the experimental data (not shown).

To improve the fit, the rate constants for the final DBP formation steps were adjusted individually, while the values for the steps forming the reaction intermediates were held at the values obtained for the optimization of Br-free waters. The final values used in the simulations are shown in Table 5.4. Figure 5.9 shows a comparison between the experimental data and the simulations for $CHCl_3$, TCAA, and the two brominated DBP species that were present at the highest concentrations in each sample. (These brominated species were $CHBrCl_2$ and $CHBr_2Cl$ for addition of 0.1, 0.2, or 0.4 mg/L Br^-, and $CHBrCl_2$ and $CHBr_3$ for addition of 1.0 mg/L Br^-).

In general, the model fit the experimental data for the brominated DBPs reasonably well. However, the effect of the presence of Br^- on the formation of $CHCl_3$ and TCAA was overestimated in all cases. That is, since the rate constants were calibrated using the data for addition of 0.4 mg/L Br^-, the model predicted (correctly) that more $CHCl_3$ and TCAA would form at lower Br^- additions, and less of these DBPs would form at higher Br^- additions, but in all cases, the model predicted a much larger change than was observed. The most likely explanation for this discrepancy is that the model incorporates an assumption that all precursor sites can be either brominated or chlorinated, whereas the results suggested that some sites that are available for bromination are not available for chlorination. Put another way, the model presumes that every bromination reaction eliminates one possible chlorination reaction, whereas the data suggest that some bromination reactions can occur without interfering with chlorination. Further study of the extent to which bromination and chlorination compete for the same reactive sites is needed to improve this aspect of the modeling.

When the model parameters determined in the LW water with added Br^- were used to simulate DBP formation in the other waters with naturally high Br^- concentrations, the fit was often not very good. The fit could be improved by adjusting the rate constants for the final DBP formation step, as was done in fitting the LW data, but the deviation between the model and experimental data remained significant at very short reaction times. Therefore, a third type of reactive site in the NOM molecules was postulated to account for very rapid formation of DBPs. This site was presumed to form small amounts of all DBPs essentially instantaneously, so that some concentration of each DBP was present immediately after the solution was dosed with the oxidant.

The concentration of each DBP that was formed instantaneously was selected as the minimum concentration of that DBP detected in the experiment. The calibration for each water was then completed by assigning the reactions that generated halogenated intermediates the same rate constants as had been used to model the halogenation of LW water, but selecting rate constants for the final step for formation of the DBPs and the values of *a* and *R* to optimize the fit between the model and the experiments. The modeling parameters from this effort are summarized in Table 5.4, and the simulated and measured concentrations of $CHCl_3$, MCAA,

Table 5.4
Optimized reaction rate constants for final DBP formation step for chlorination of waters containing significant concentrations of bromide

	Nueces	Jordan	Ft. Worth	Hillsborough*
k_{Cl1}	3.4×10^{-7}	1.4×10^{-5}	2.5×10^{-5}	2.5×10^{-7}
k_{CHCl3}	1.8×10^{4}	0.48	0.34	17
k_{TCAA}	1.9×10^{4}	0.28	0.26	11
k_{DCAA}	4.3×10^{-3}	9.0×10^{-4}	6.4×10^{-4}	8.8×10^{-4}
k_{MCAA}	0.20	0.059	0.060	4.2×10^{-2}
R	6.7	18	14	2
a	9.6	23	20	1

*water diluted 6× prior to analysis

DCAA, TCAA, and the two brominated DBPs that were detected at the highest concentration in each water are shown in Figure 5.10.

In all these cases, the model fit the data quite well for all the DBPs shown. Furthermore, the values obtained for the branching ratio, shown in Table 5.4, are comparable with those reported in the literature. For example, Cowman and Singer (1996) reported that the best-fit R values for HAAs and THMs were 10 and 20, respectively, while Nokes, Fenton, and Randall (1999) determined that R varied from 4 to 15, with an average close to 9. The R value determined in this study for the Nueces River is very close to those reported in the above publications, while the preference for bromination is somewhat higher, but nevertheless of the same magnitude, for NOM from the Jordan River and Forth Worth water. Although the estimate of R was much lower in the Hillsborough water, the Br^- concentration in that water was also quite low, making the estimate less reliable.

The values of a (the ratio of the rate constants for the release of the DBP from the halogenated intermediates via reaction with $Cl_{2,tot}$ vs. $Br_{2,tot}$) were always substantially >1, indicating a higher reactivity of the intermediates with HOBr than with HOCl in the final DBP formation step. In general, values of a were also larger than the corresponding values of R.

If the final DBP release step involves oxidation of the intermediate, the relatively high values of a would be counter-intuitive, because, thermodynamically, HOCl is a more potent oxidant than HOBr. One plausible interpretation is that the breakdown of the trihalogenated intermediates to release THMs and HAAs is not strictly, or even primarily, oxidative. For instance, the breakdown of the trihalogenated intermediates may proceed through the generation of an additional, short-lived intermediate whose formation requires incorporation of a fourth halogen atom into the trihalogenated substrate. The ensuing rapid hydrolysis of that transient species could then liberate both a THM or HAA and a halogen ion (Cl^- or Br^-). Other explanations are also possible, of course.

SUMMARY

This chapter summarizes the development and testing of a conventional, mechanistic kinetic model to characterize THM and HAA formation during chlorination of typical drinking water sources. The model envisions halogenation of NOM molecules to occur stepwise, up to a point where three halogen atoms are attached to a carbon atom in the molecule. After each halogenation step, thc molecule can be cleaved to release an HAA, or, once a site has three halogen atoms attached, it might release a THM or undergo another reaction to form an unidentified DBP. All reactions are envisioned to be first-order with respect to the organic species and the halogen, and second-order overall.

To fit the experimental data for waters containing very low concentrations of bromide, it was necessary to postulate that at least two kinetically distinct precursor sites exist in NOM. With that assumption, the data for formation of the major DBPs ($CHCl_3$ and TCAA) could be modeled well for a wide range of Cl_2:DOC ratios, pH values, and water sources, even when the fitting parameters were constrained by certain assumptions about the similarity of the reaction pathways for the two sites. The data for MCAA and DCAA were not fit as well by the model, because the generation pattern for these two DBPs consisted of extremely rapid formation of the compounds, followed by very little additional formation. Postulating that a third type of site exists as the source of the DBPs that appear extremely rapidly in these systems would allow MCAA and DCAA formation to be modeled very well. For all the DBPs modeled, the rate constants suggested that the initial attack of a site in an NOM molecule by an HOCl or OCl^- molecule is rate-limiting.

Modeling of DBP formation in waters containing a substantial concentration of bromide ion is a good deal more complex, due to the multiple branching reactions that can occur and lead to formation of mixed-halogen DBPs. Some success was achieved in modeling of such waters by incorporating two additional parameters into the model to characterize the relative reactivities of the organic species with free bromine as opposed to free chlorine. In addition, it was necessary to invoke the presence of a third type of site in NOM to account for the very rapid formation of some DBPs, as suggested above. Even with those additional parameters, DBP formation in a given water with widely varying Br^- concentrations could not be modeled by a single set of parameters—such an approach consistently overestimated the effect of changing the Br^- concentration on DBP speciation. Nevertheless, if a water has a reasonably stable Br^- concentration, the model could be used to describe the kinetics and speciation of DBP formation.

While additional work can certainly improve the correspondence between the model and experimental observations of DBP formation, it is worth noting that very few mechanistic models exist that even attempt to characterize formation of the full suite of THMs and HAAs in a consistent, scientifically defensible framework, and those that do exist (e.g., the model of McClellan et al. [2000]) use many more fitting parameters than the model proposed here. In that sense, the proposed model represents a significant advance toward the goal of characterizing DBP formation kinetics within a tractable and realistic framework.

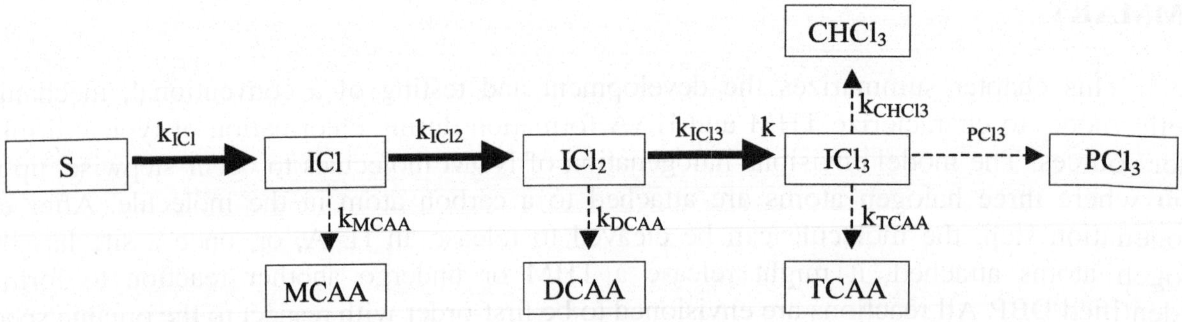

Figure 5.1 Proposed reaction pathways for THM and HAA formation in systems with negligible Br^-. Both the fast- and slow-reacting sites were assumed to follow this pathway.

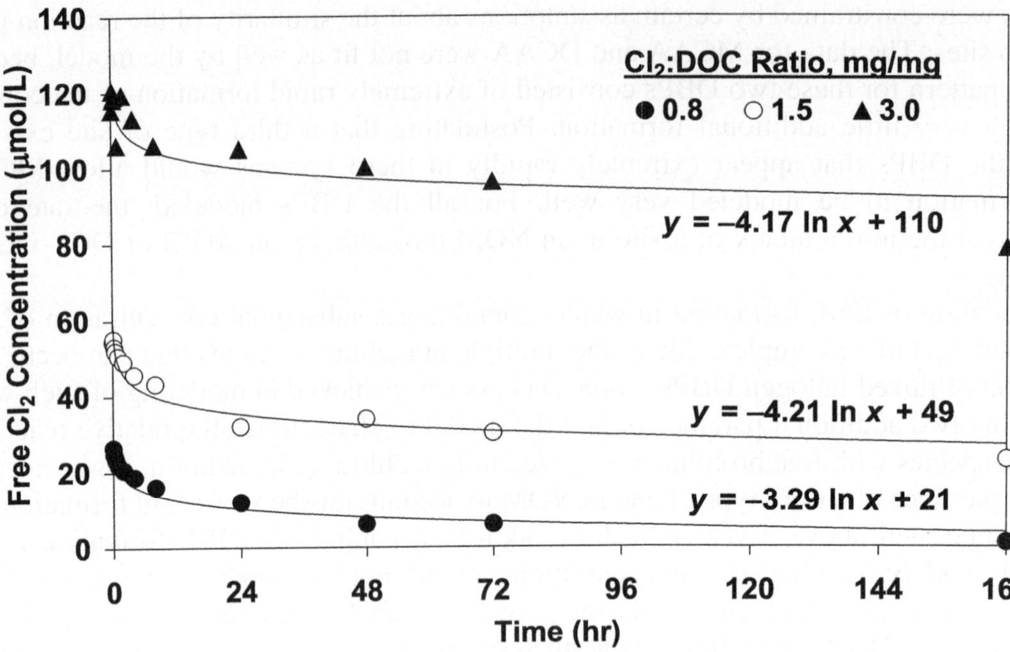

Figure 5.2 Fit of model equation to experimental data for Cl_2 decay kinetics in Lake Washington water at various Cl_2:DOC ratios

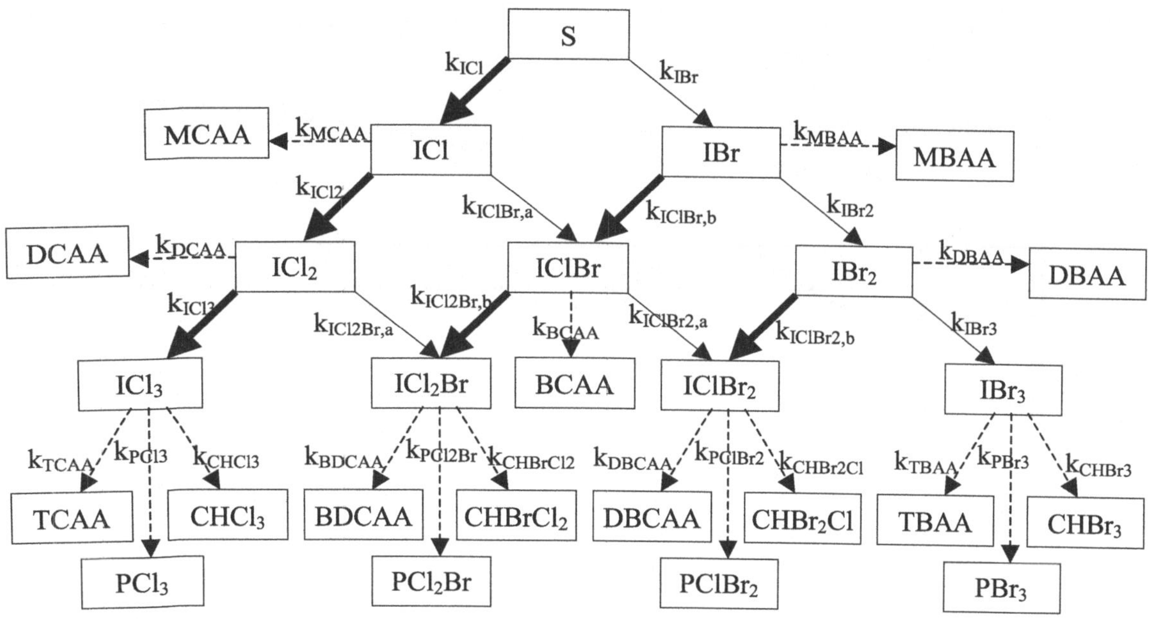

Figure 5.3 Proposed reaction scheme for waters containing significant Br^- concentrations

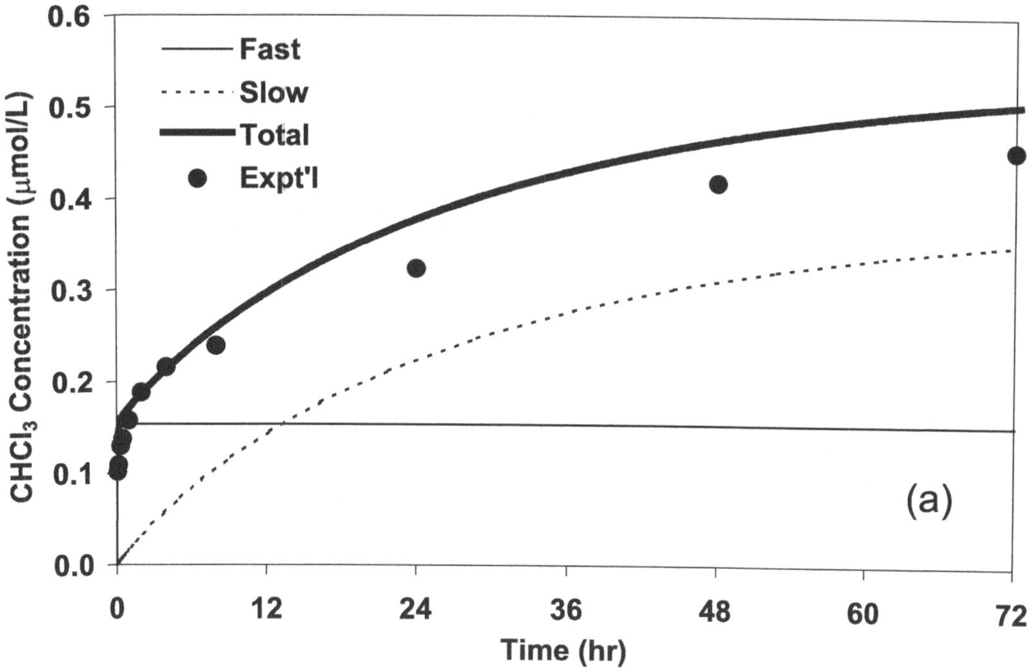

Figure 5.4a Comparison of model and experimental results for DBP formation in Lake Washington water at a dosing ratio of $Cl_2/DOC = 1.5$. The curves show the DBP formation predicted to occur from fast-reacting, slow-reacting, and total sites, respectively. (a) $CHCl_3$.

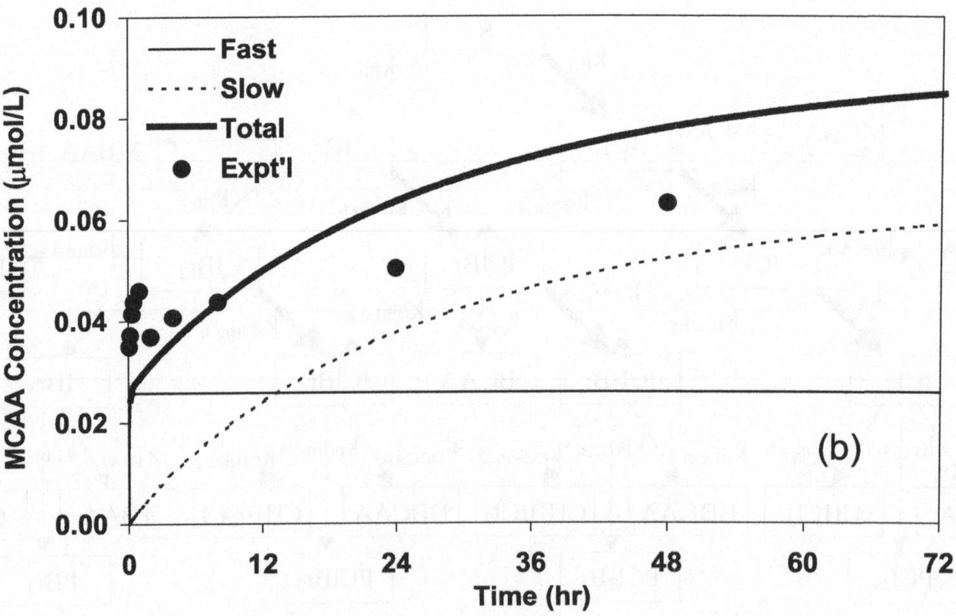

Figure 5.4b Comparison of model and experimental results for DBP formation in Lake Washington water at a dosing ratio of Cl_2/DOC = 1.5. The curves show the DBP formation predicted to occur from fast-reacting, slow-reacting, and total sites, respectively. (b) MCAA.

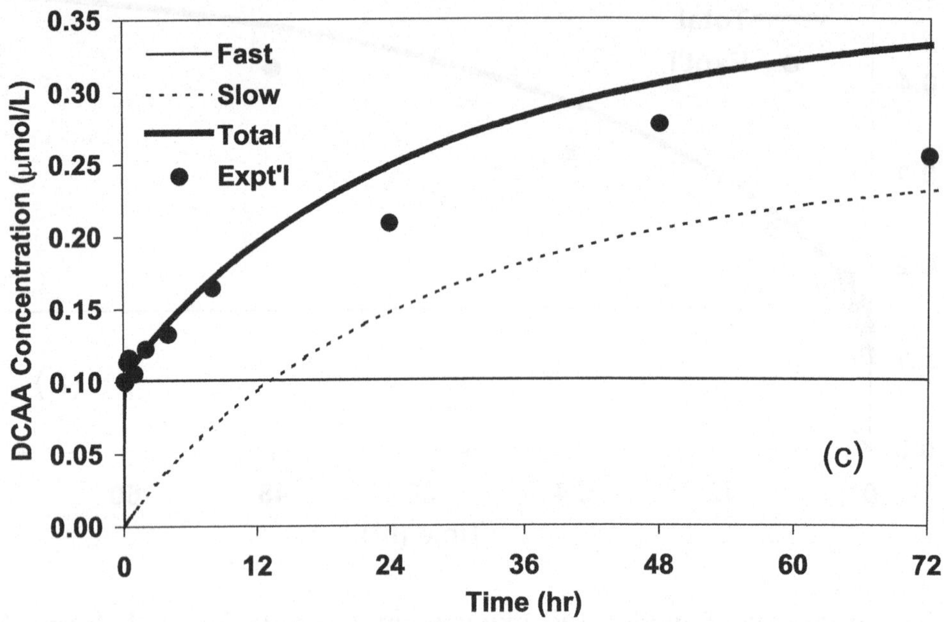

Figure 5.4c Comparison of model and experimental results for DBP formation in Lake Washington water at a dosing ratio of Cl_2/DOC = 1.5. The curves show the DBP formation predicted to occur from fast-reacting, slow-reacting, and total sites, respectively. (c) DCAA.

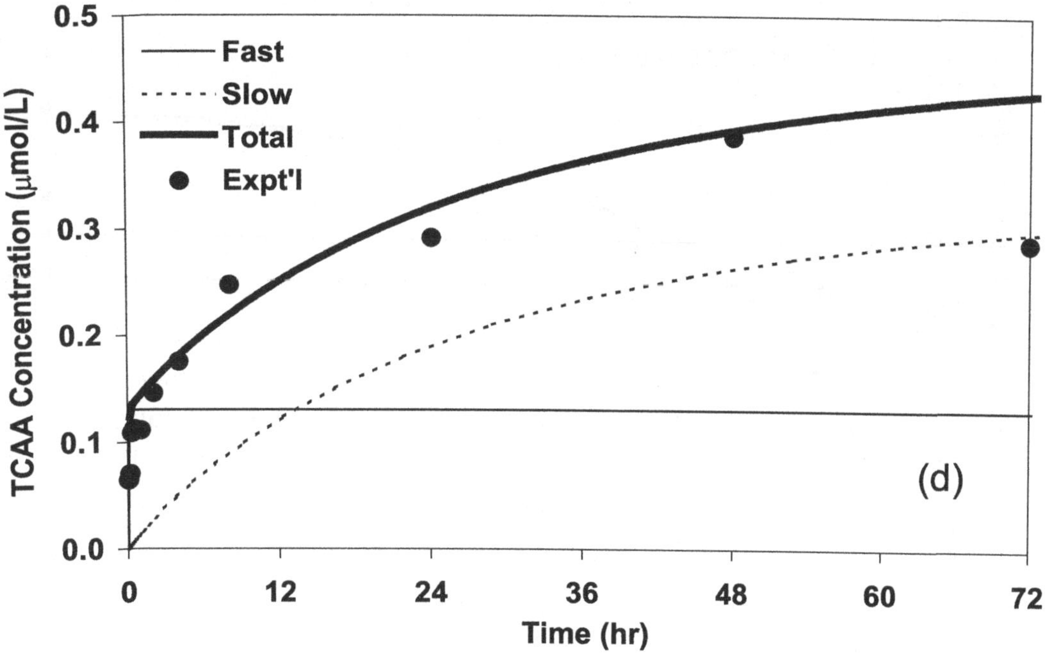

Figure 5.4d Comparison of model and experimental results for DBP formation in Lake Washington water at a dosing ratio of Cl_2/DOC = 1.5. The curves show the DBP formation predicted to occur from fast-reacting, slow-reacting, and total sites, respectively. (d) TCAA.

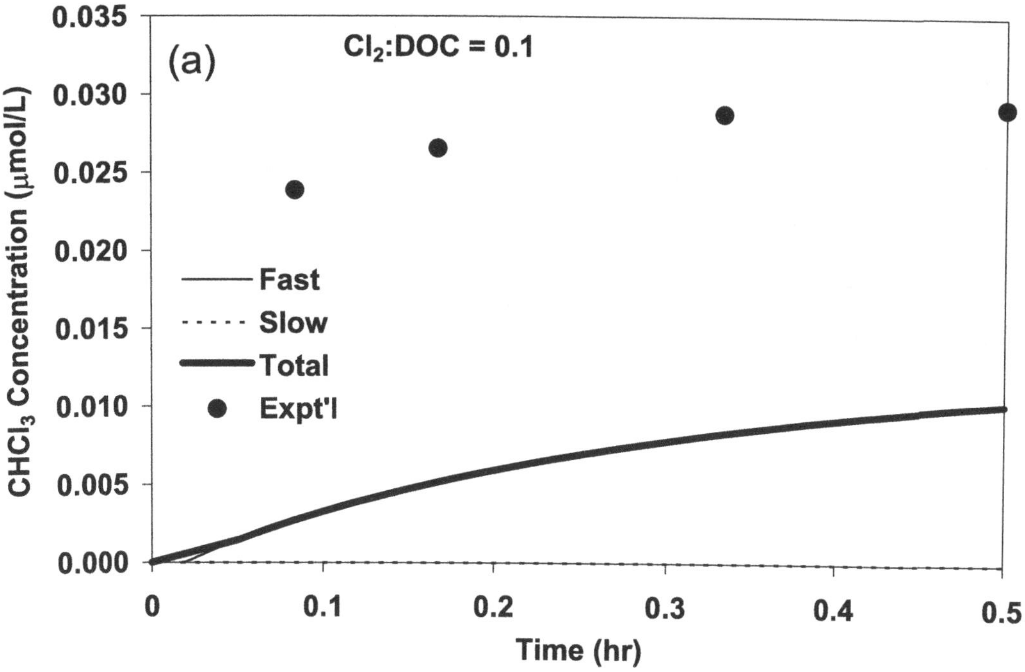

Figure 5.5a Comparison of model and experimental results for $CHCl_3$ formation in Lake Washington water at various Cl_2:DOC ratios. (a) 0.1 mg/mg.

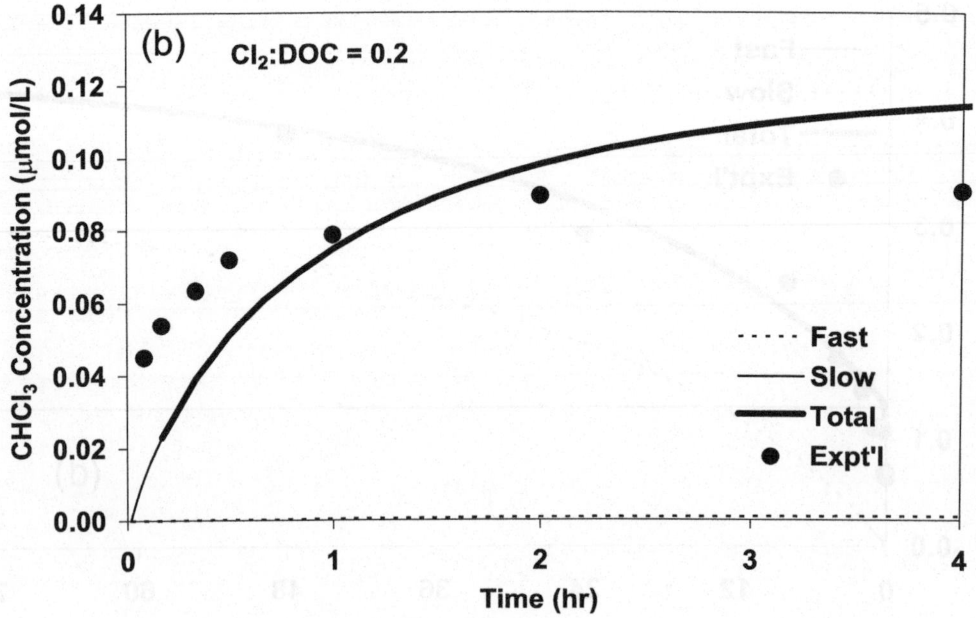

Figure 5.5b Comparison of model and experimental results for $CHCl_3$ formation in Lake Washington water at various Cl_2:DOC ratios. (b) 0.2 mg/mg.

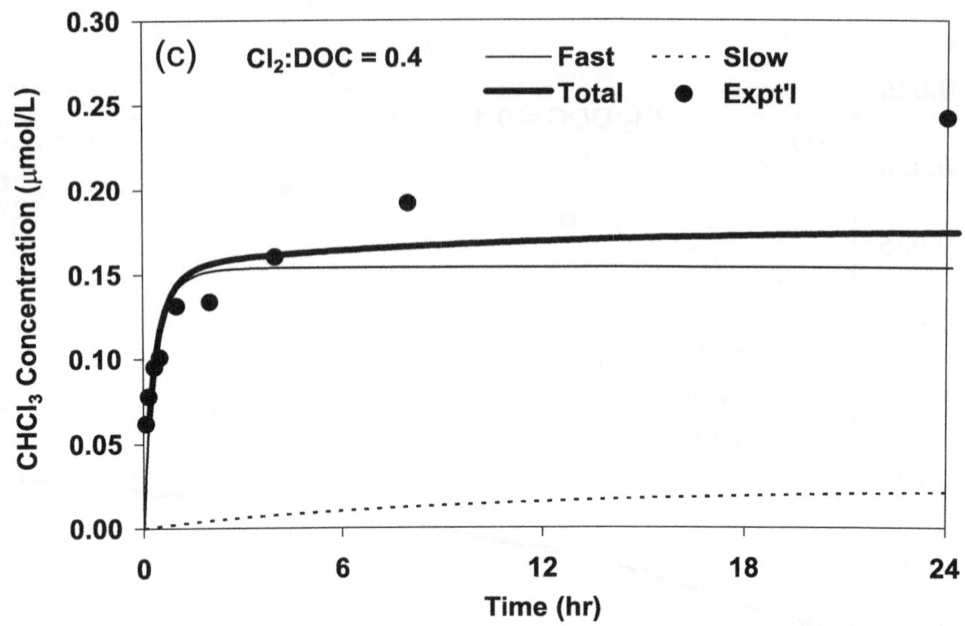

Figure 5.5c Comparison of model and experimental results for $CHCl_3$ formation in Lake Washington water at various Cl_2:DOC ratios. (c) 0.4 mg/mg.

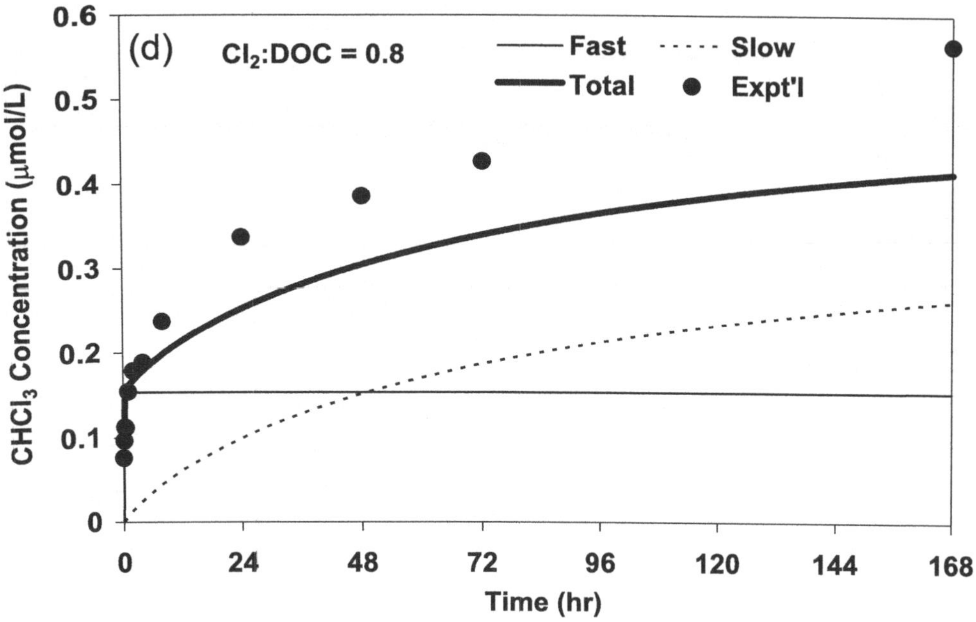

Figure 5.5d Comparison of model and experimental results for $CHCl_3$ formation in Lake Washington water at various Cl_2:DOC ratios. (d) 0.8 mg/mg.

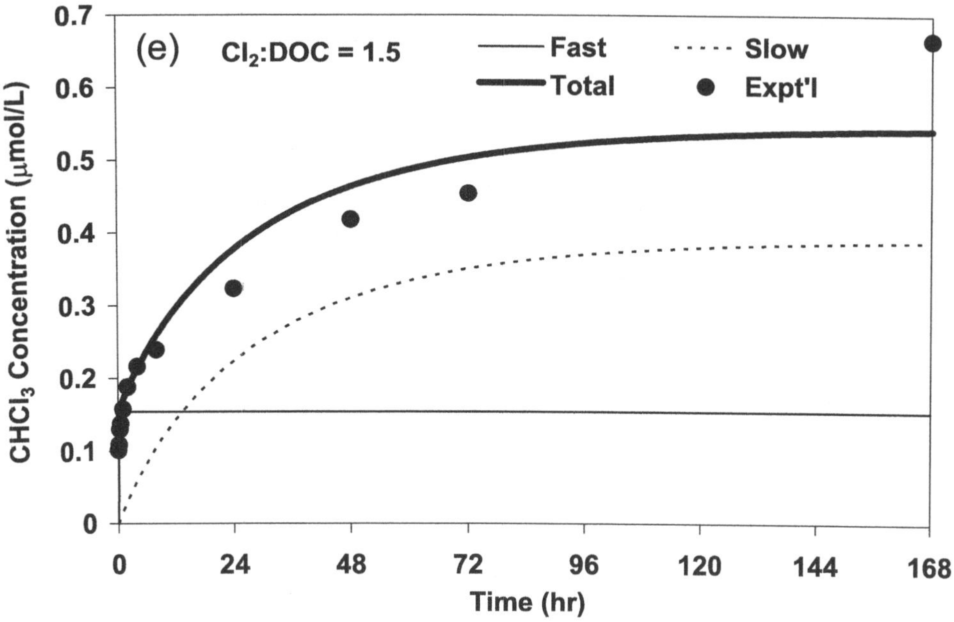

Figure 5.5e Comparison of model and experimental results for $CHCl_3$ formation in Lake Washington water at various Cl_2:DOC ratios. (e) 1.5 mg/mg.

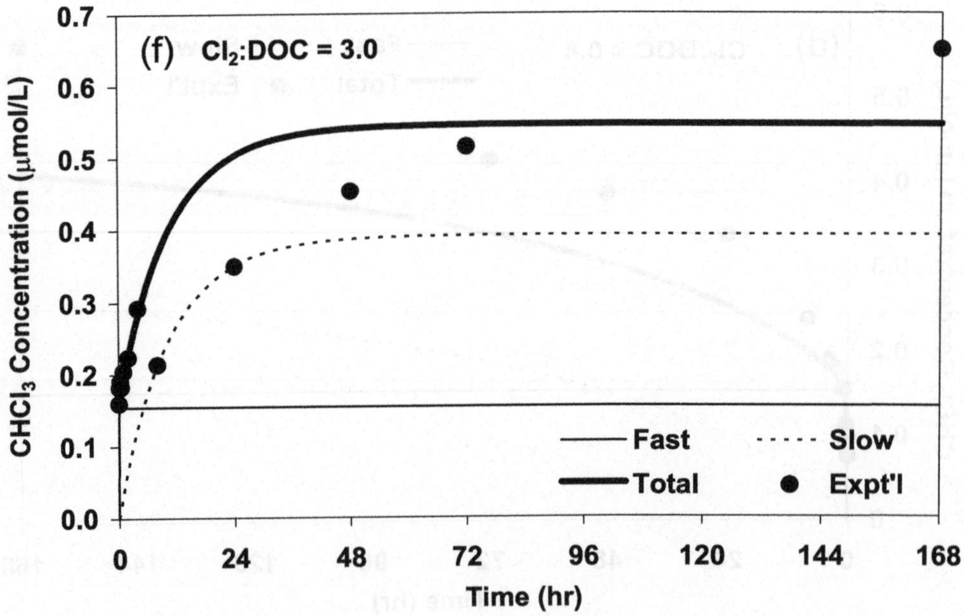

Figure 5.5f Comparison of model and experimental results for $CHCl_3$ formation in Lake Washington water at various Cl_2:DOC ratios. (f) 3.0 mg/mg.

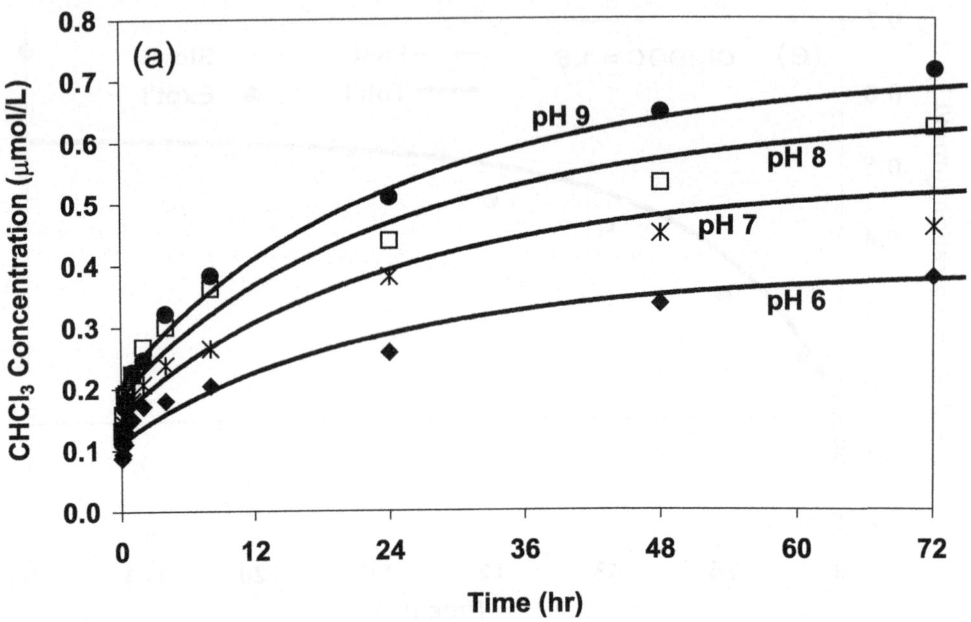

Figure 5.6a Comparison of model and experimental results for $CHCl_3$ and TCAA formation in Lake Washington water at various pH values. (a) $CHCl_3$.

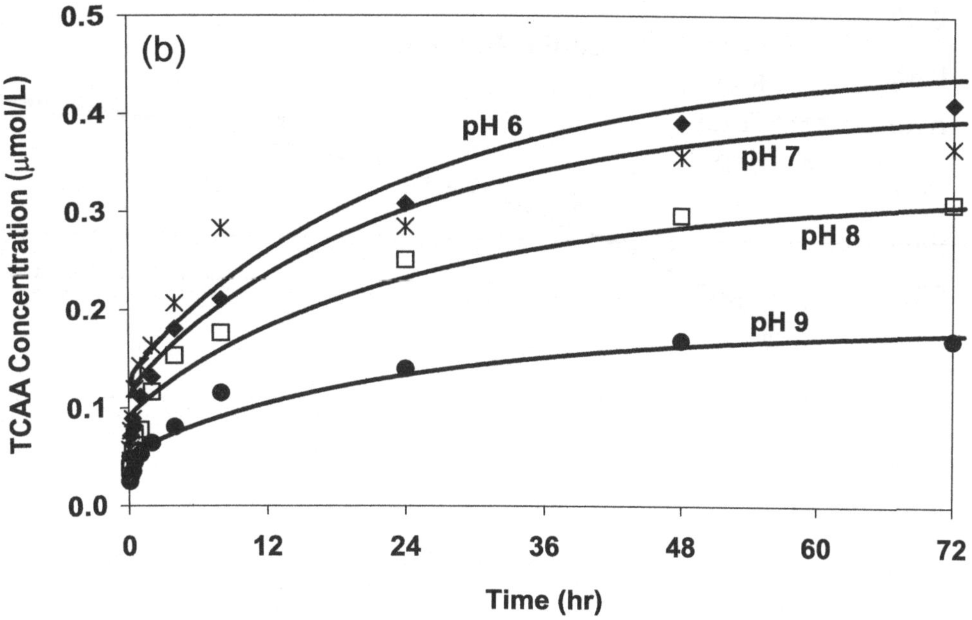

Figure 5.6b Comparison of model and experimental results for $CHCl_3$ and TCAA formation in Lake Washington water at various pH values. (b) TCAA.

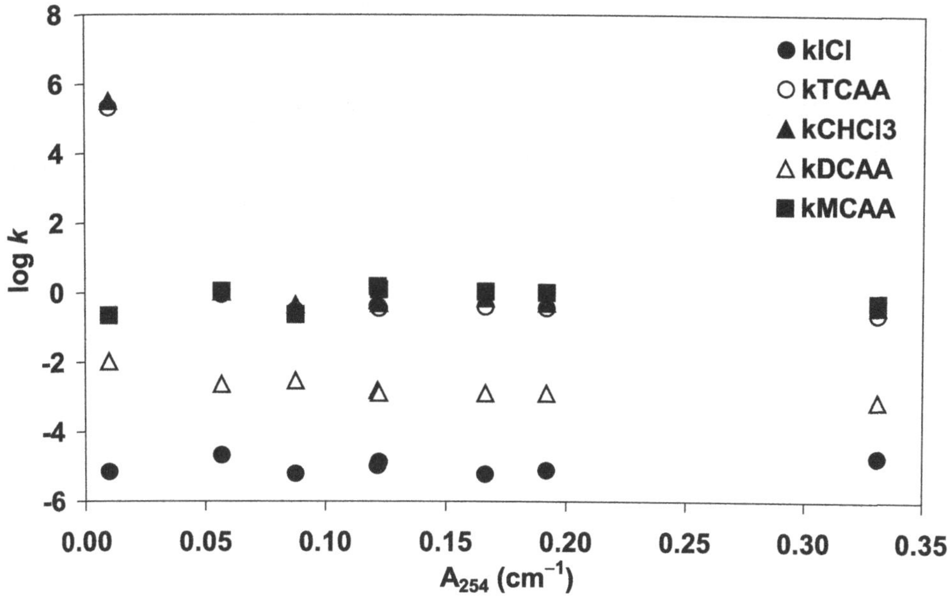

Figure 5.7 Optimized reaction rate constants as a function of A_{254} for various water sources. The water sources associated with the A_{254} values are: 0.010—Little Cottonwood; 0.057—Lake Washington; 0.088—Eagle Mt. Lake; 0.12—Florence WTP; 0.17—St. Paul WTP; 0.19—Sioux Falls WTP; 0.33—Minneapolis WTP.

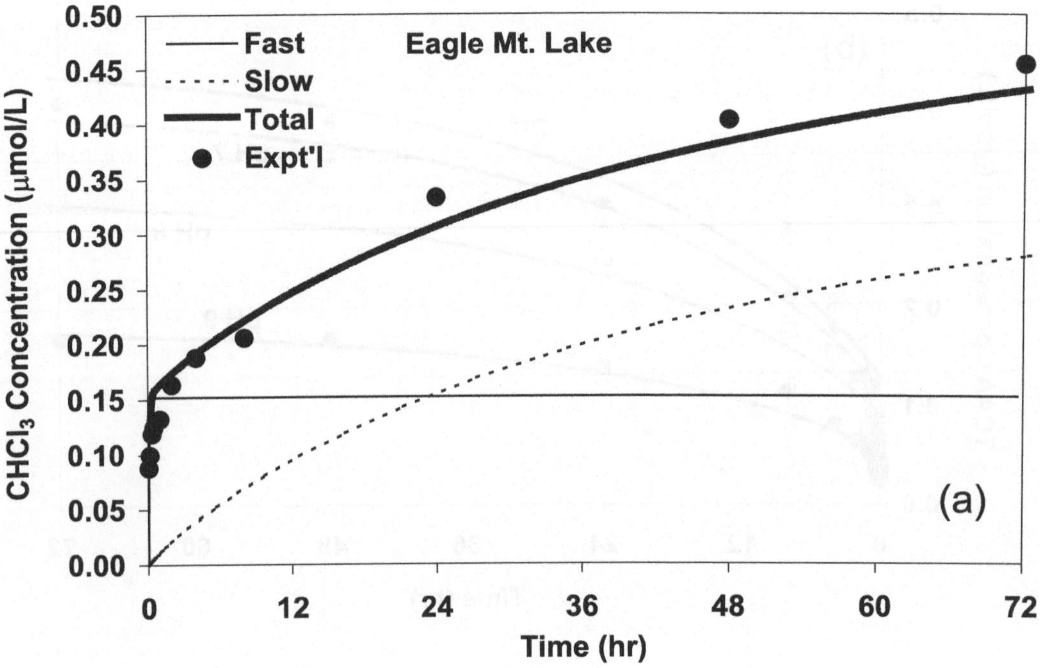

Figure 5.8a Comparison of model and experimental data for $CHCl_3$ and TCAA formation in various waters. (a) $CHCl_3$ in Eagle Mt. Lake.

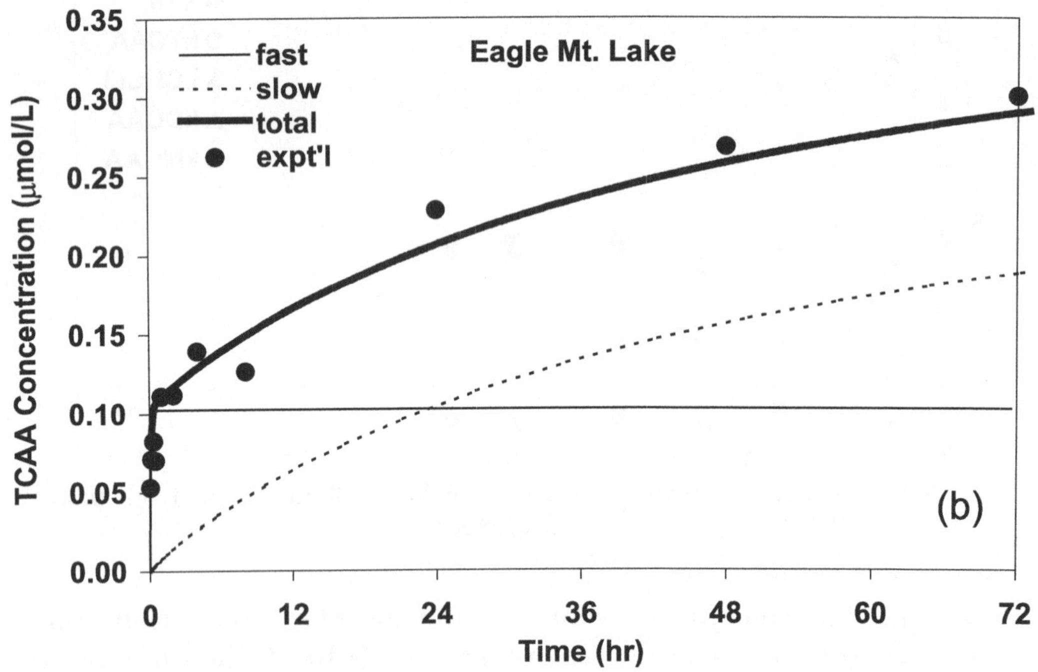

Figure 5.8b Comparison of model and experimental data for $CHCl_3$ and TCAA formation in various waters. (b) TCAA in Eagle Mt. Lake.

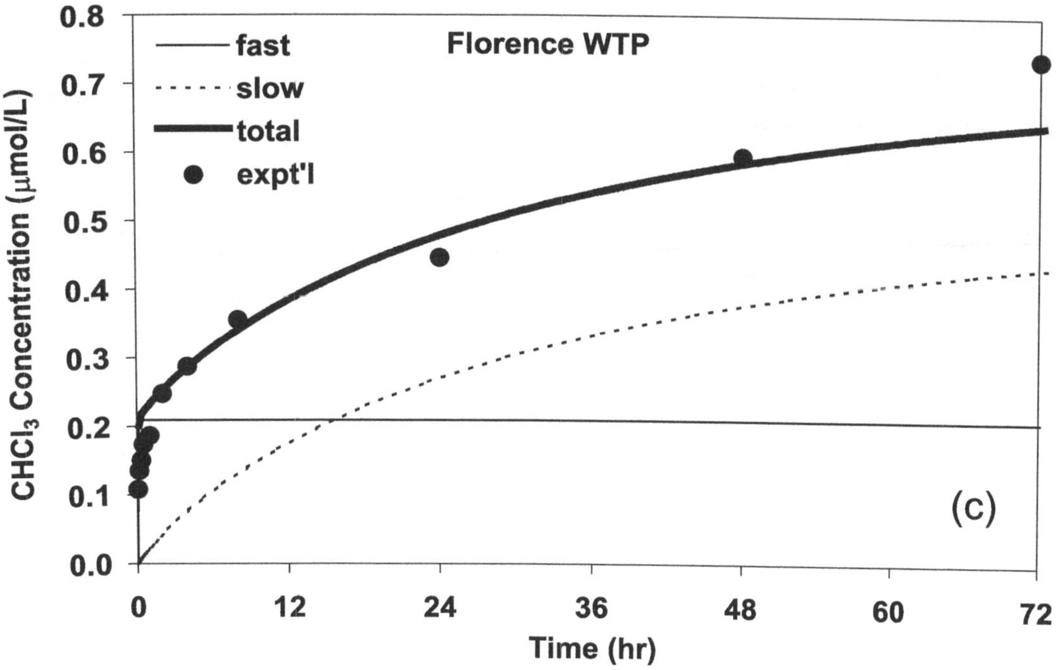

Figure 5.8c Comparison of model and experimental data for $CHCl_3$ and TCAA formation in various waters. (c) $CHCl_3$ in Florence WTP water.

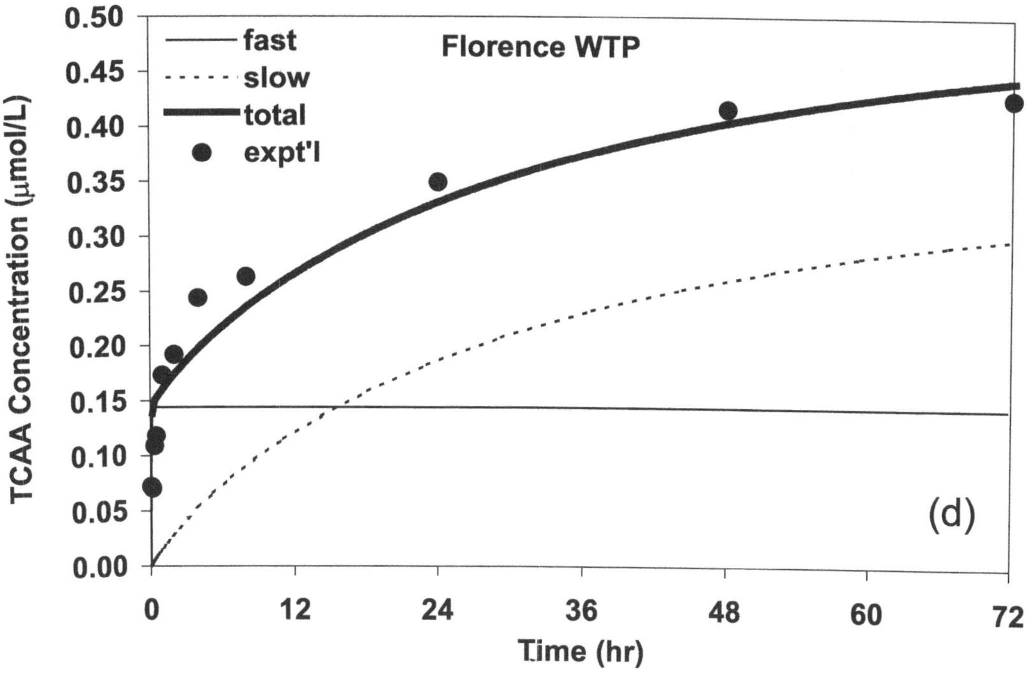

Figure 5.8d Comparison of model and experimental data for $CHCl_3$ and TCAA formation in various waters. (d) TCAA in Florence WTP water.

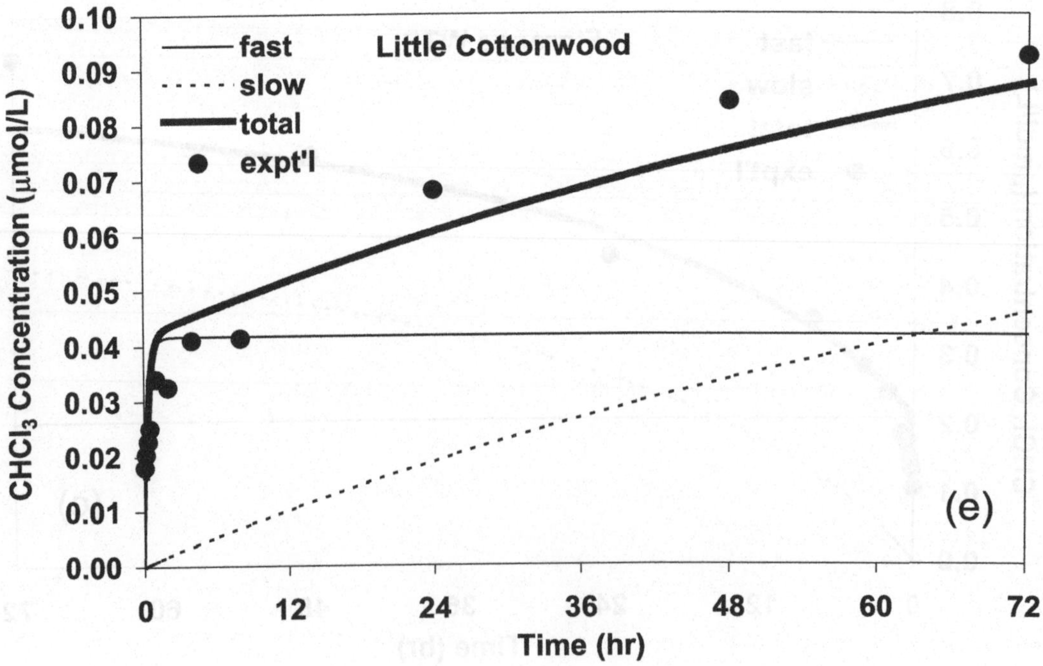

Figure 5.8e Comparison of model and experimental data for $CHCl_3$ and TCAA formation in various waters. (e) $CHCl_3$ in Little Cottonwood Creek water.

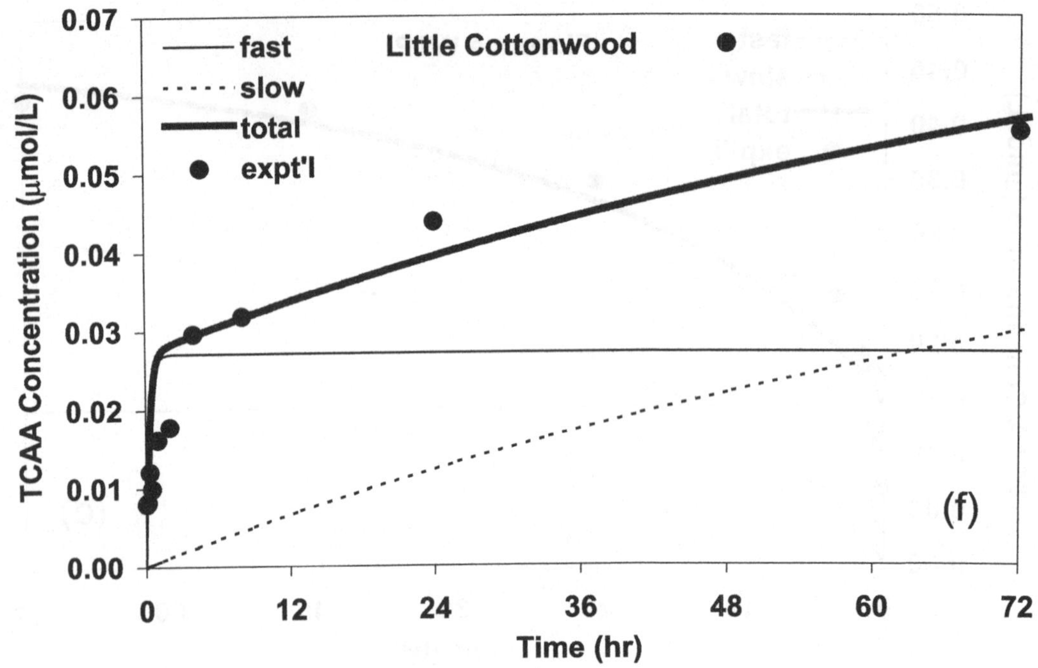

Figure 5.8f Comparison of model and experimental data for $CHCl_3$ and TCAA formation in various waters. (f) TCAA in Little Cottonwood Creek water.

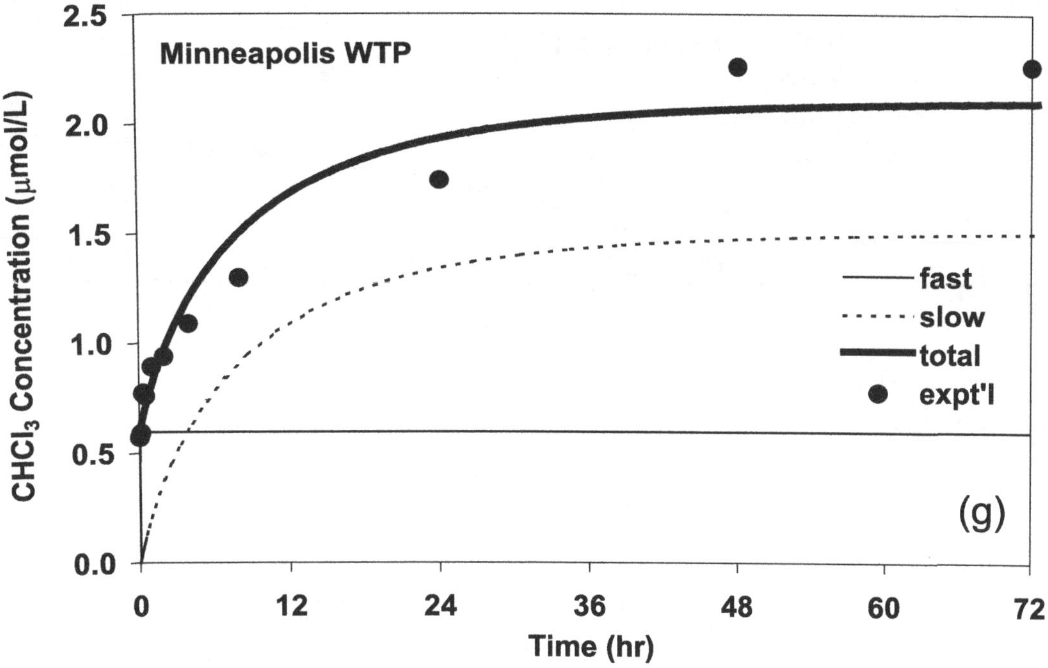

Figure 5.8g Comparison of model and experimental data for $CHCl_3$ and TCAA formation in various waters. (g) $CHCl_3$ in Minneapolis WTP water.

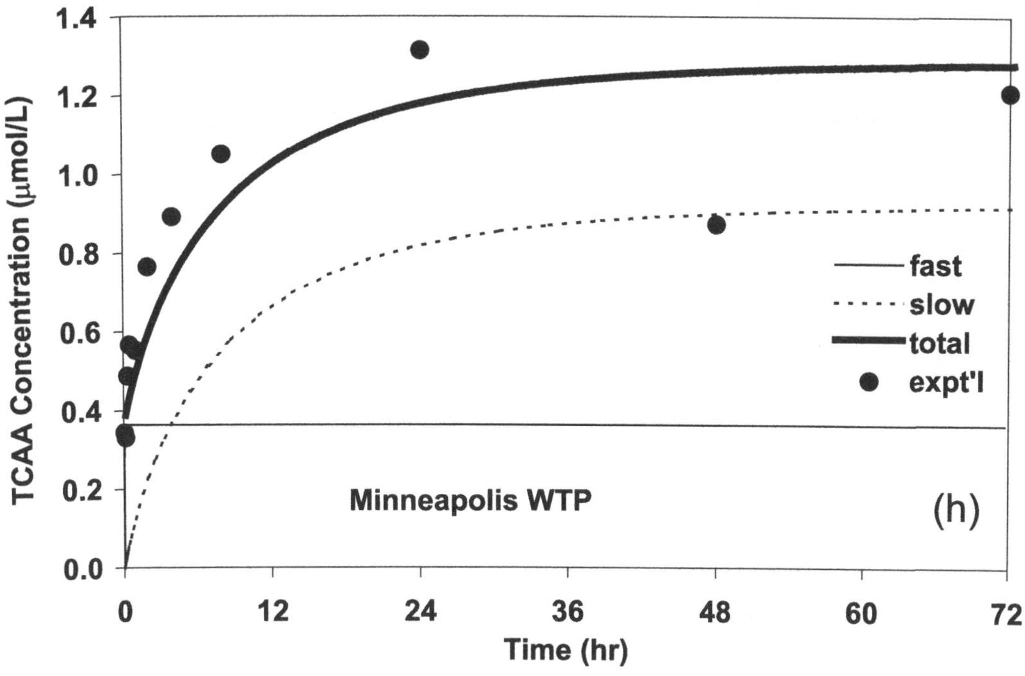

Figure 5.8h Comparison of model and experimental data for $CHCl_3$ and TCAA formation in various waters. (h) TCAA in Minneapolis WTP water.

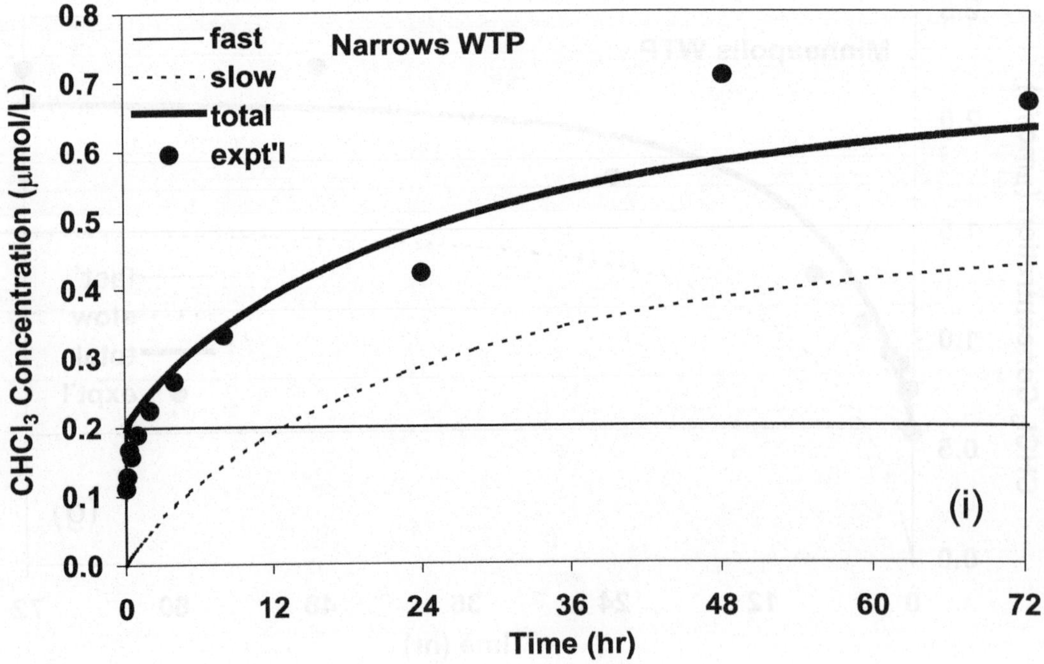

Figure 5.8i Comparison of model and experimental data for $CHCl_3$ and TCAA formation in various waters. (i) $CHCl_3$ in Narrows WTP water.

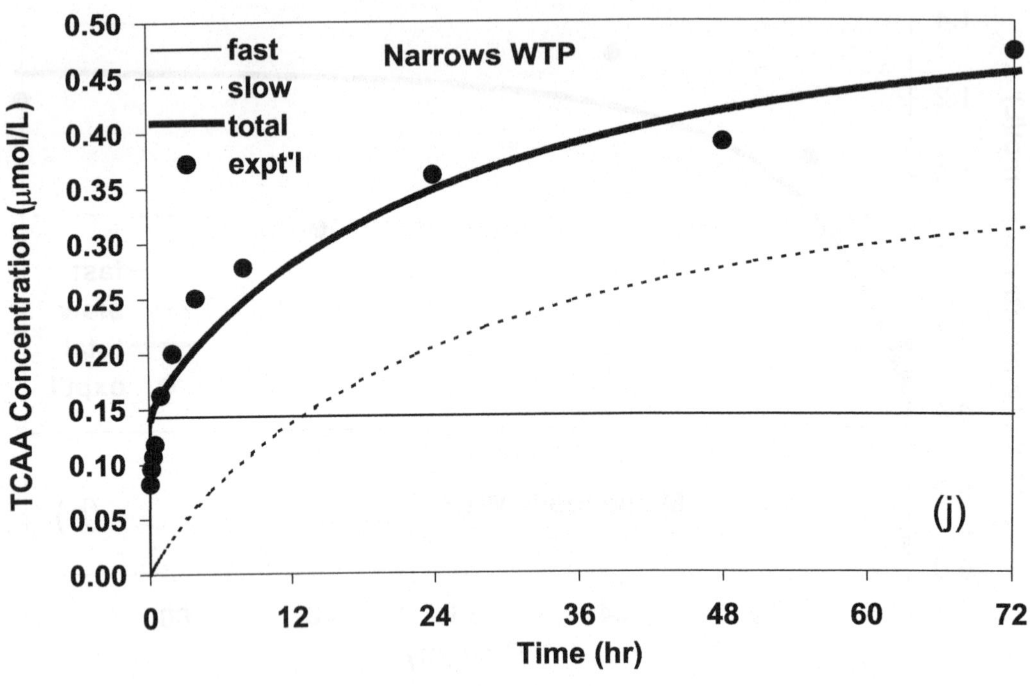

Figure 5.8j Comparison of model and experimental data for $CHCl_3$ and TCAA formation in various waters. (j) TCAA in Narrows WTP water.

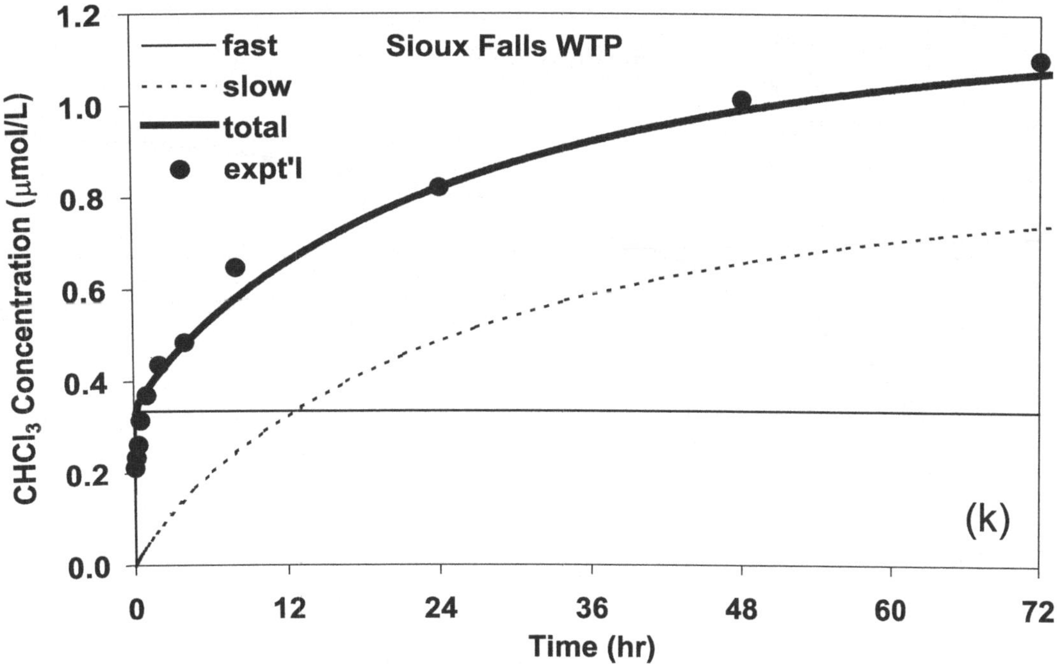

Figure 5.8k Comparison of model and experimental data for $CHCl_3$ and TCAA formation in various waters. (k) $CHCl_3$ in Sioux Falls WTP water.

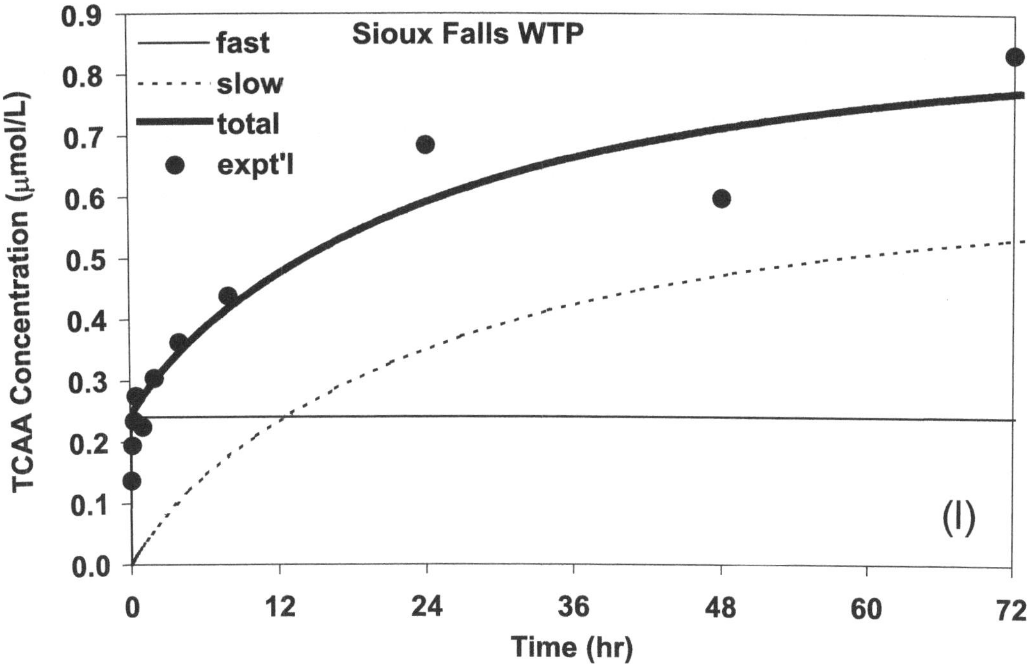

Figure 5.8l Comparison of model and experimental data for $CHCl_3$ and TCAA formation in various waters. (l) TCAA in Narrows WTP water.

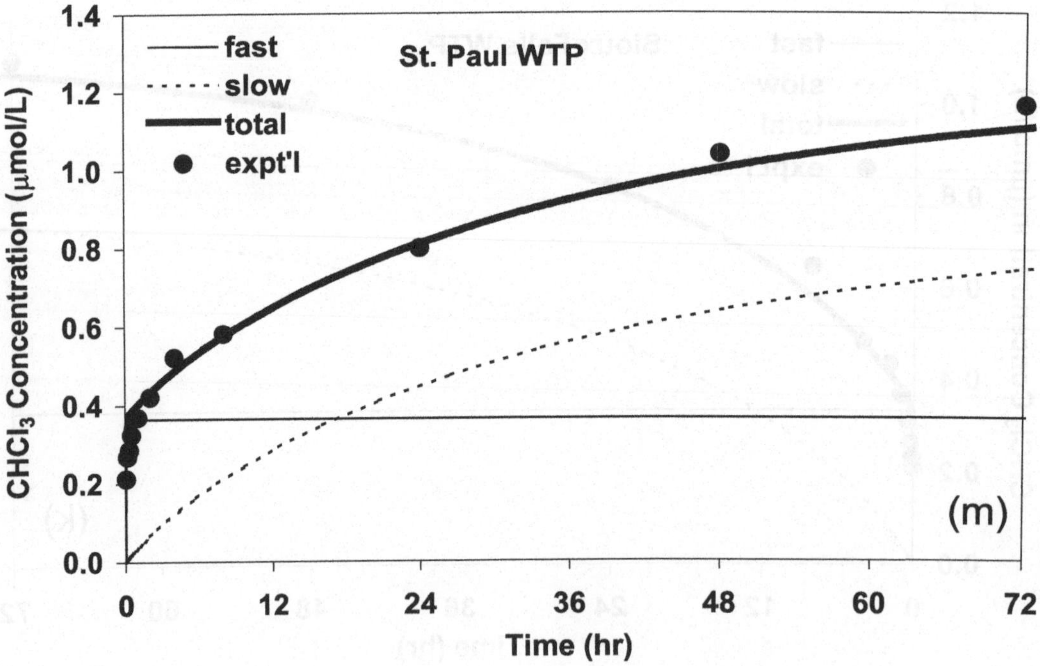

Figure 5.8m Comparison of model and experimental data for $CHCl_3$ and TCAA formation in various waters. (m) $CHCl_3$ in St. Paul WTP water.

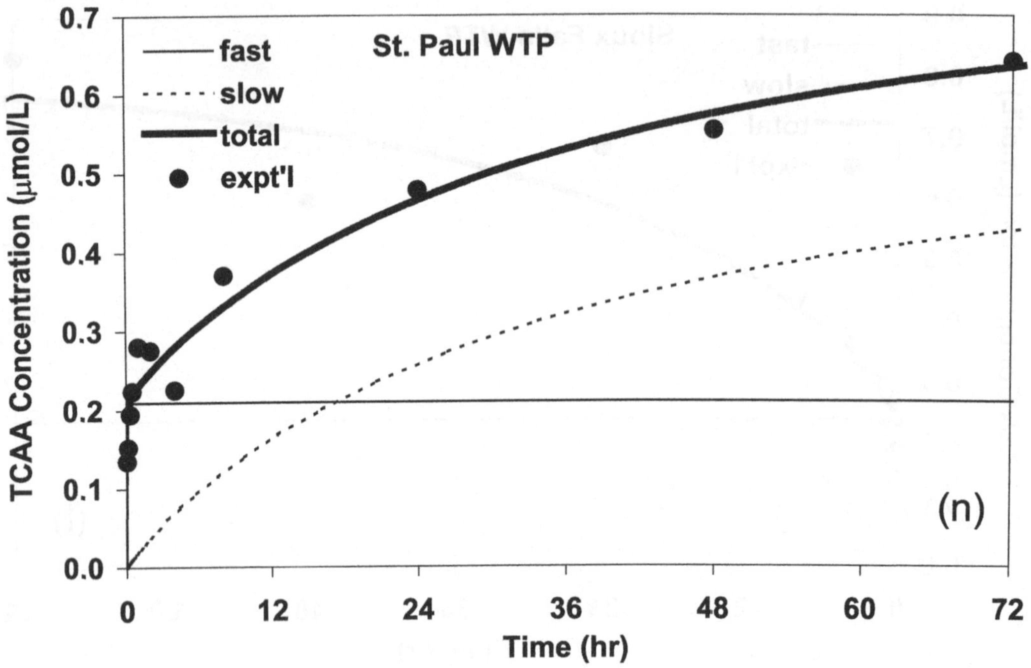

Figure 5.8n Comparison of model and experimental data for $CHCl_3$ and TCAA formation in various waters. (n) TCAA in St. Paul WTP water.

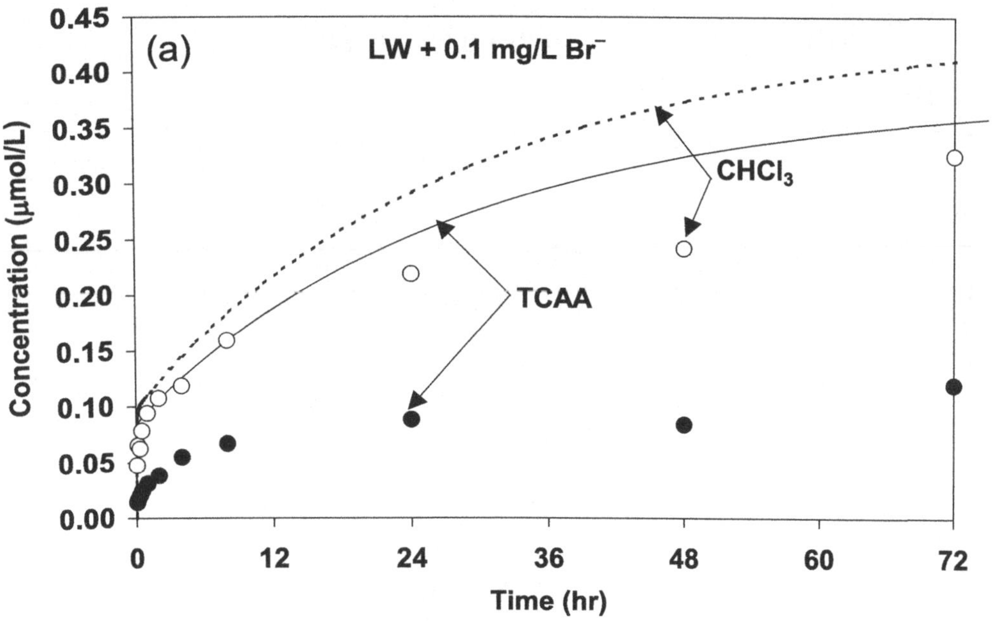

Figure 5.9a Comparison of experimental data (points) and model simulations (lines) for $CHCl_3$, TCAA, and the two brominated DBPs found at the highest concentrations in chlorinated Lake Washington water with various Br^- additions. (a) $CHCl_3$ and TCAA, 0.1 mg/L Br^- added.

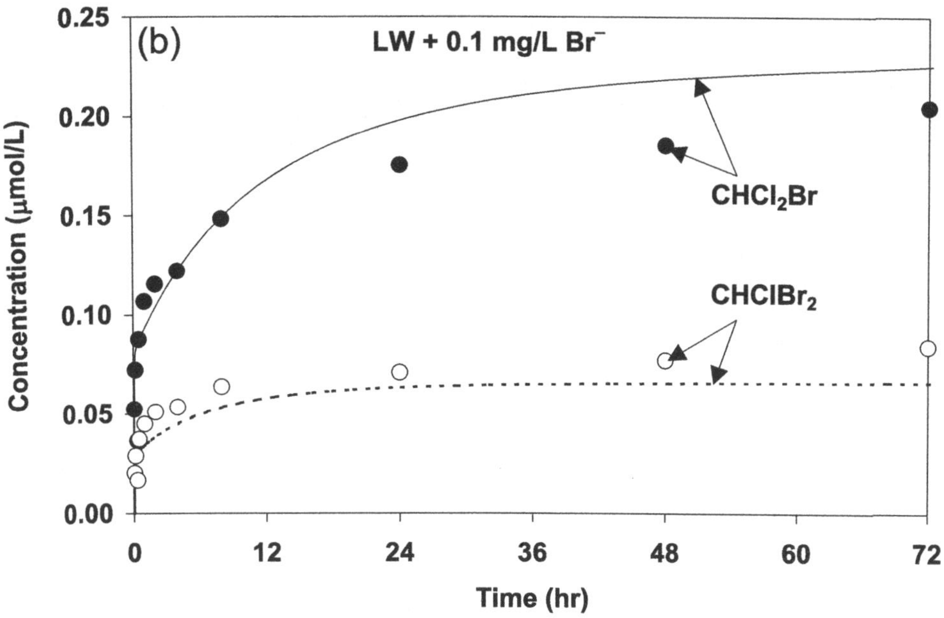

Figure 5.9b Comparison of experimental data (points) and model simulations (lines) for $CHCl_3$, TCAA, and the two brominated DBPs found at the highest concentrations in chlorinated Lake Washington water with various Br^- additions. (b) $CHCl_2Br$ and $CHClBr_2$, 0.1 mg/L Br^- added.

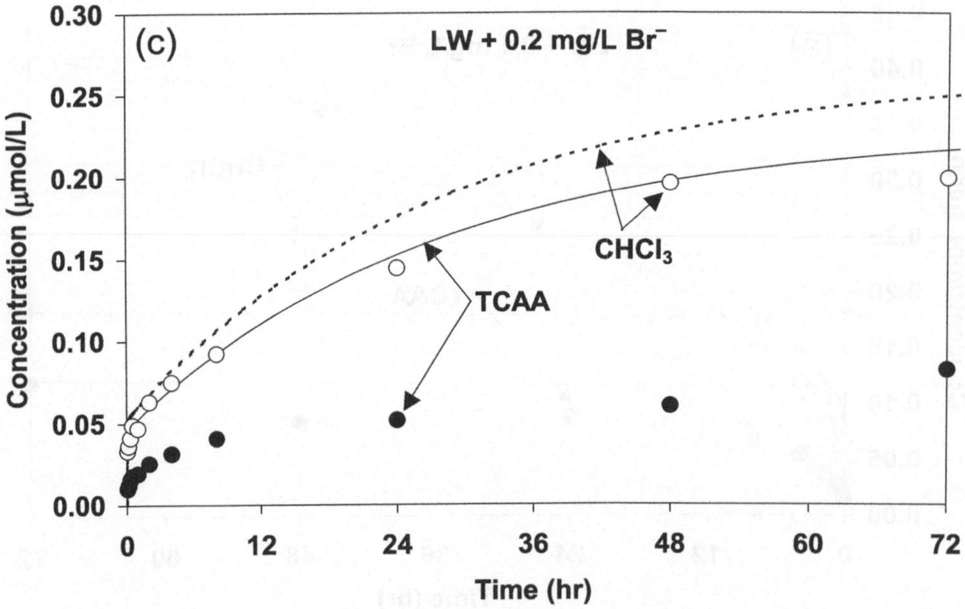

Figure 5.9c Comparison of experimental data (points) and model simulations (lines) for $CHCl_3$, TCAA, and the two brominated DBPs found at the highest concentrations in chlorinated Lake Washington water with various Br^- additions. (c) $CHCl_3$ and TCAA, 0.2 mg/L Br^- added.

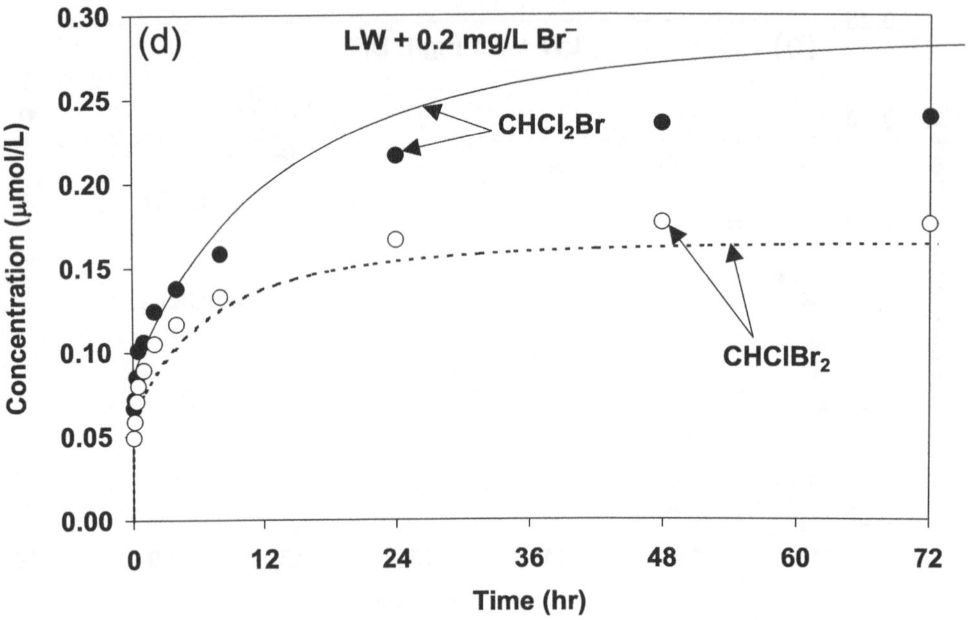

Figure 5.9d Comparison of experimental data (points) and model simulations (lines) for $CHCl_3$, TCAA, and the two brominated DBPs found at the highest concentrations in chlorinated Lake Washington water with various Br^- additions. (d) $CHCl_2Br$ and $CHClBr_2$, 0.2 mg/L Br^- added.

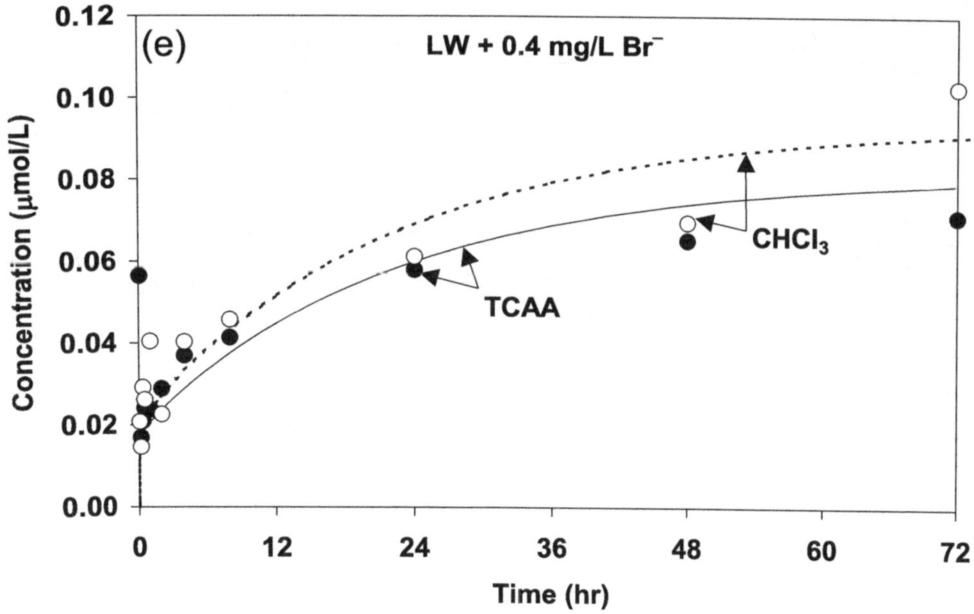

Figure 5.9e Comparison of experimental data (points) and model simulations (lines) for $CHCl_3$, TCAA, and the two brominated DBPs found at the highest concentrations in chlorinated Lake Washington water with various Br^- additions. (e) $CHCl_3$ and TCAA, 0.4 mg/L Br^- added.

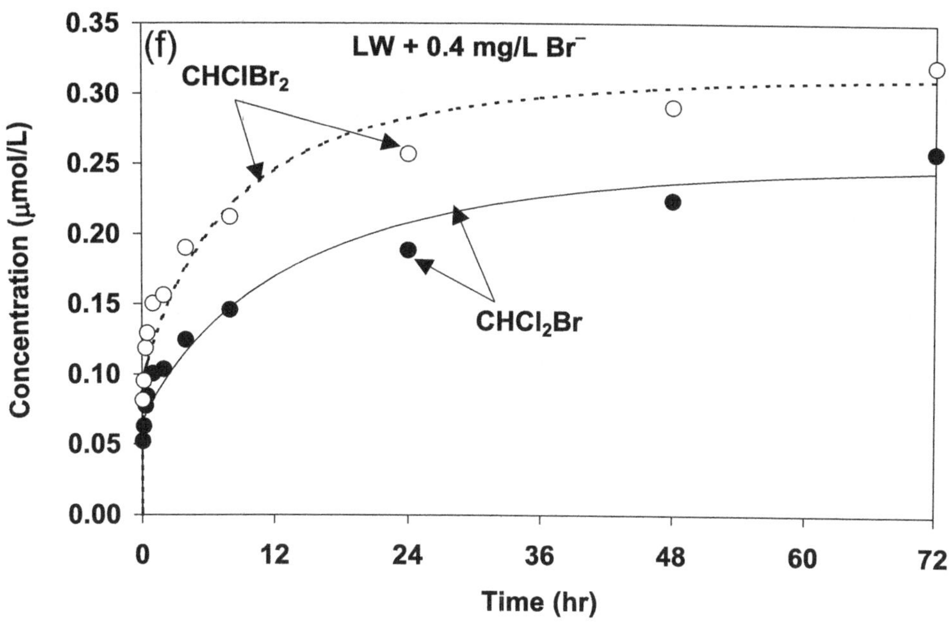

Figure 5.9f Comparison of experimental data (points) and model simulations (lines) for $CHCl_3$, TCAA, and the two brominated DBPs found at the highest concentrations in chlorinated Lake Washington water with various Br^- additions. (f) $CHCl_2Br$ and $CHClBr_2$, 0.4 mg/L Br^- added.

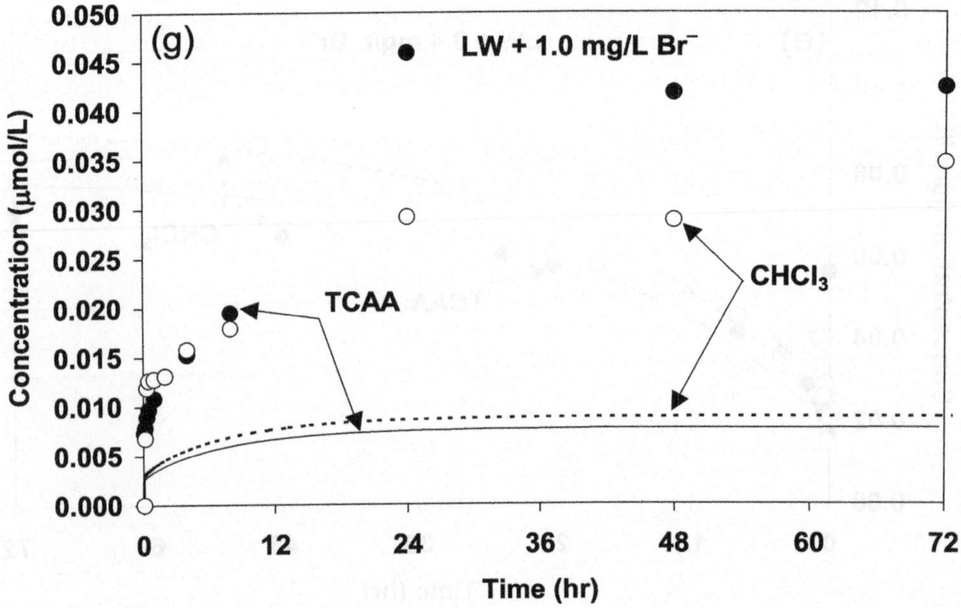

Figure 5.9g Comparison of experimental data (points) and model simulations (lines) for $CHCl_3$, TCAA, and the two brominated DBPs found at the highest concentrations in chlorinated Lake Washington water with various Br^- additions. (g) $CHCl_3$ and TCAA, 1.0 mg/L Br^- added.

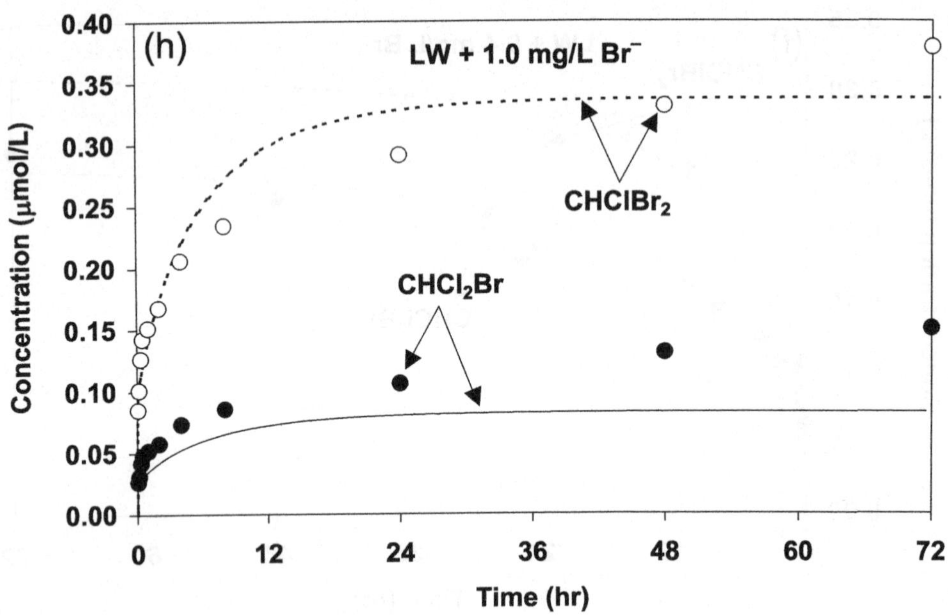

Figure 5.9h Comparison of experimental data (points) and model simulations (lines) for $CHCl_3$, TCAA, and the two brominated DBPs found at the highest concentrations in chlorinated Lake Washington water with various Br^- additions. (h) $CHCl_2Br$ and $CHClBr_2$, 1.0 mg/L Br^- added.

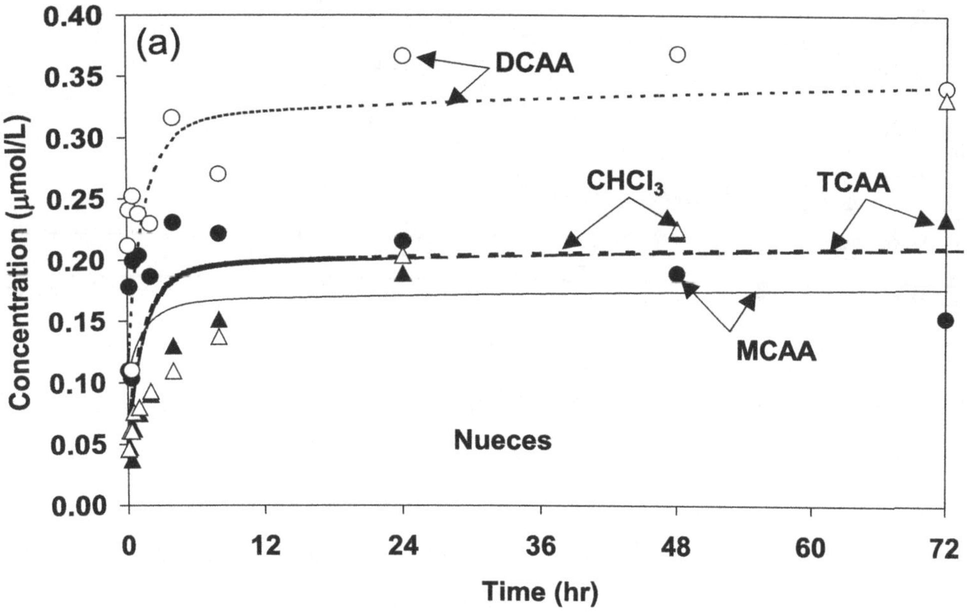

Figure 5.10a Comparison of experimental data (points) and model simulations (lines) for $CHCl_3$, HAAs, and the two brominated DBPs found at the highest concentrations in several chlorinated water sources. (a) $CHCl_3$, MCAA, DCAA, and TCAA in Nueces River water.

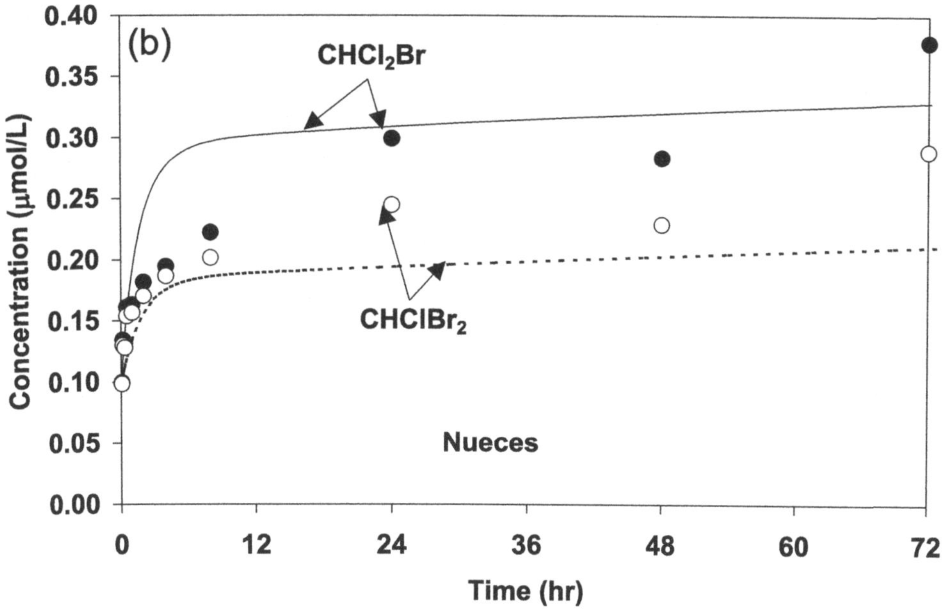

Figure 5.10b Comparison of experimental data (points) and model simulations (lines) for $CHCl_3$, HAAs, and the two brominated DBPs found at the highest concentrations in several chlorinated water sources. (b) $CHCl_2Br$ and $CHClBr_2$ in Nueces River water.

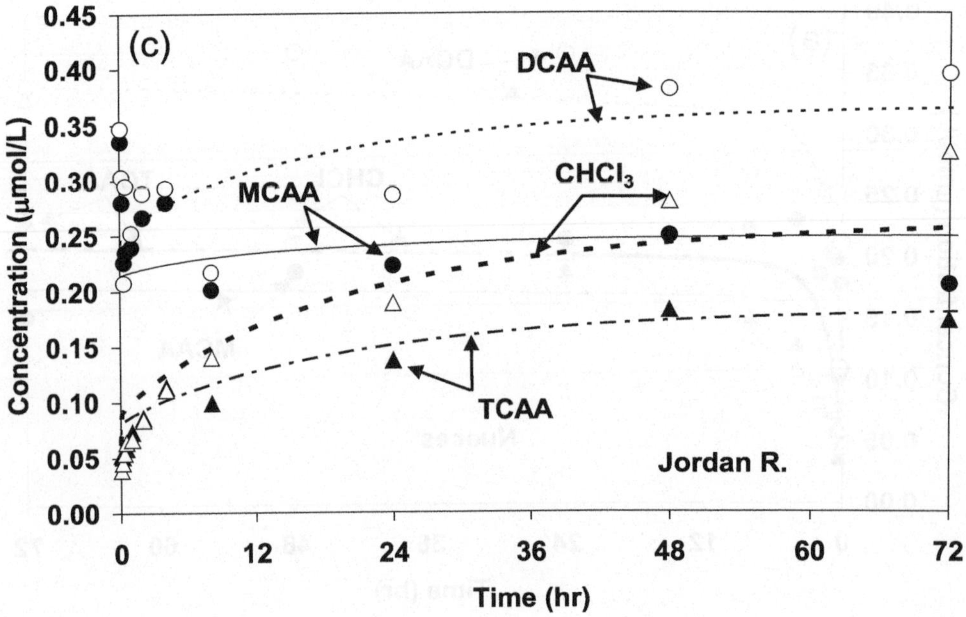

Figure 5.10c Comparison of experimental data (points) and model simulations (lines) for $CHCl_3$, HAAs, and the two brominated DBPs found at the highest concentrations in several chlorinated water sources. (c) $CHCl_3$, MCAA, DCAA, and TCAA in Jordan River water.

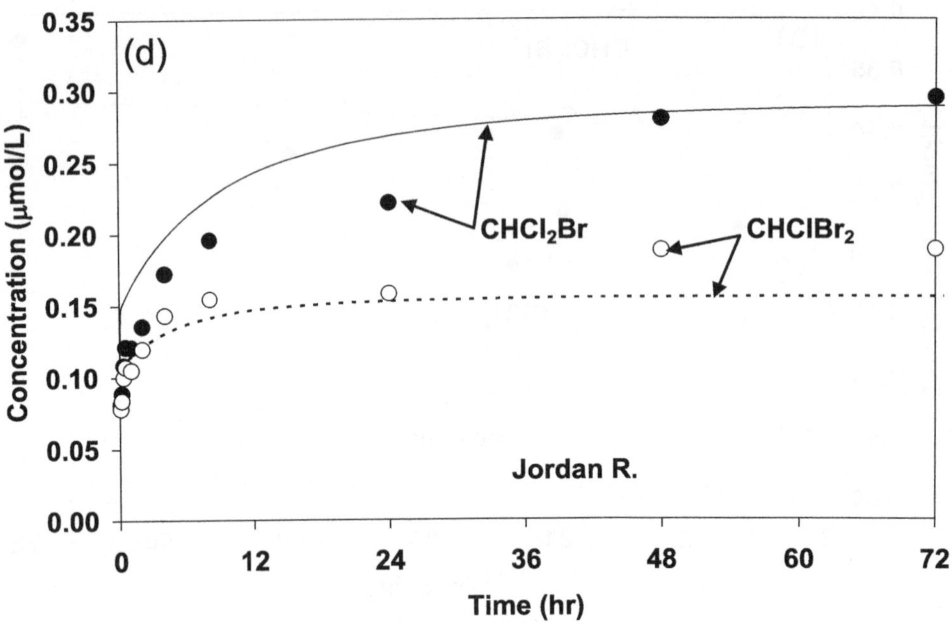

Figure 5.10d Comparison of experimental data (points) and model simulations (lines) for $CHCl_3$, HAAs, and the two brominated DBPs found at the highest concentrations in several chlorinated water sources. (d) $CHCl_2Br$ and $CHClBr_2$ in Jordan River water.

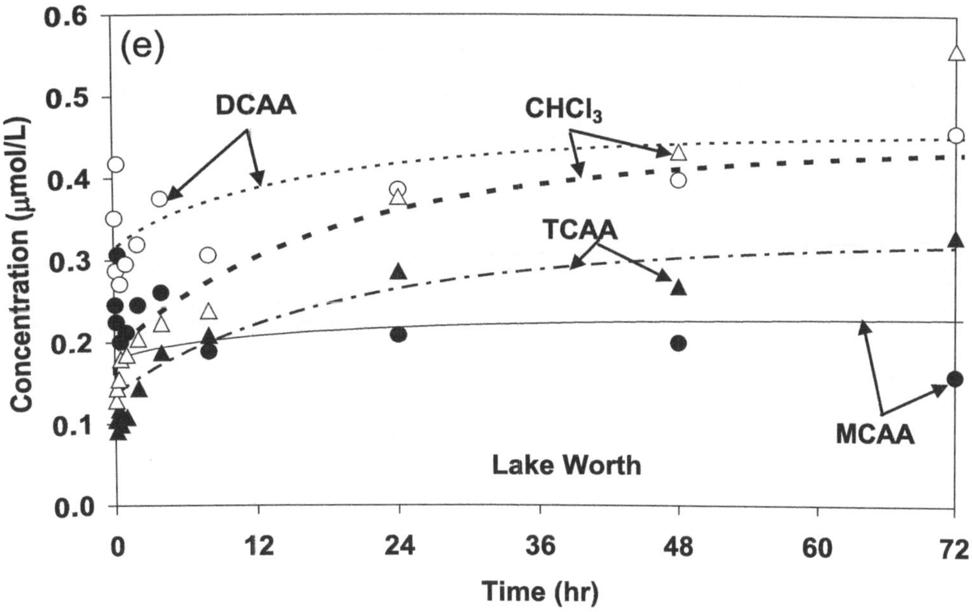

Figure 5.10e Comparison of experimental data (points) and model simulations (lines) for $CHCl_3$, HAAs, and the two brominated DBPs found at the highest concentrations in several chlorinated water sources. (e) $CHCl_3$, MCAA, DCAA, and TCAA in Lake Worth water.

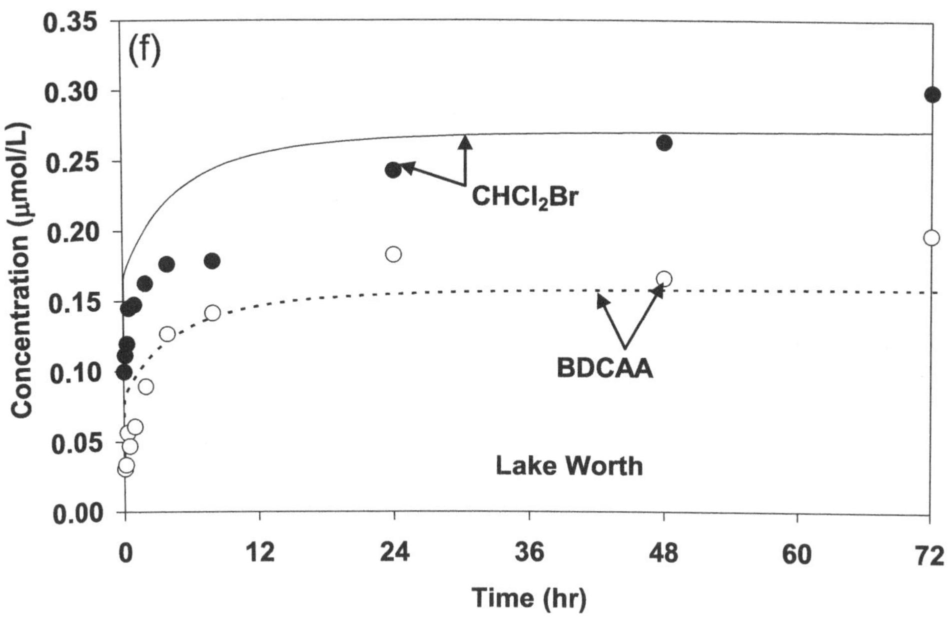

Figure 5.10f Comparison of experimental data (points) and model simulations (lines) for $CHCl_3$, HAAs, and the two brominated DBPs found at the highest concentrations in several chlorinated water sources. (f) $CHCl_2Br$ and BDCAA in Lake Worth water.

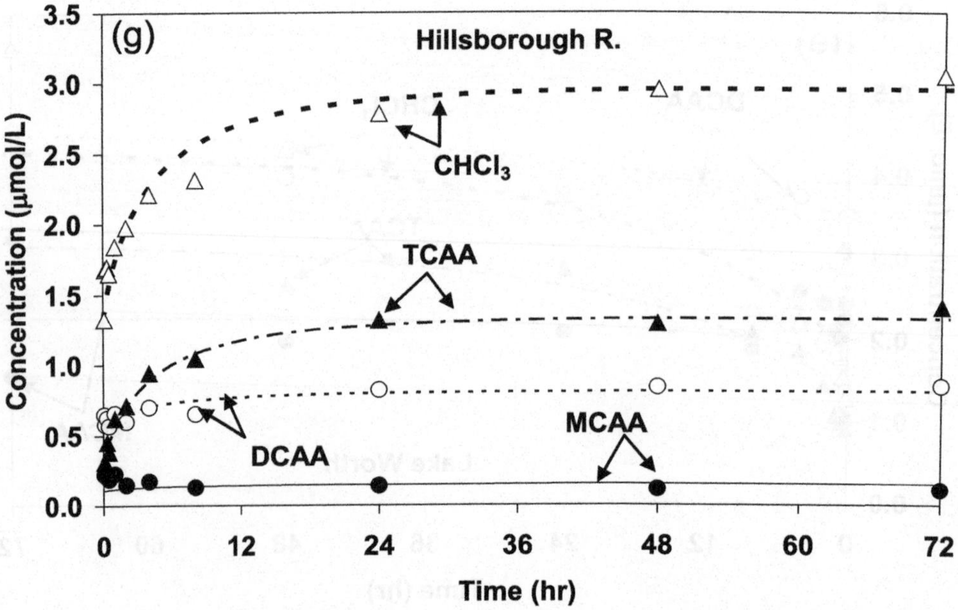

Figure 5.10g Comparison of experimental data (points) and model simulations (lines) for $CHCl_3$, HAAs, and the two brominated DBPs found at the highest concentrations in several chlorinated water sources. (g) $CHCl_3$, MCAA, DCAA, and TCAA in Hillsborough River water.

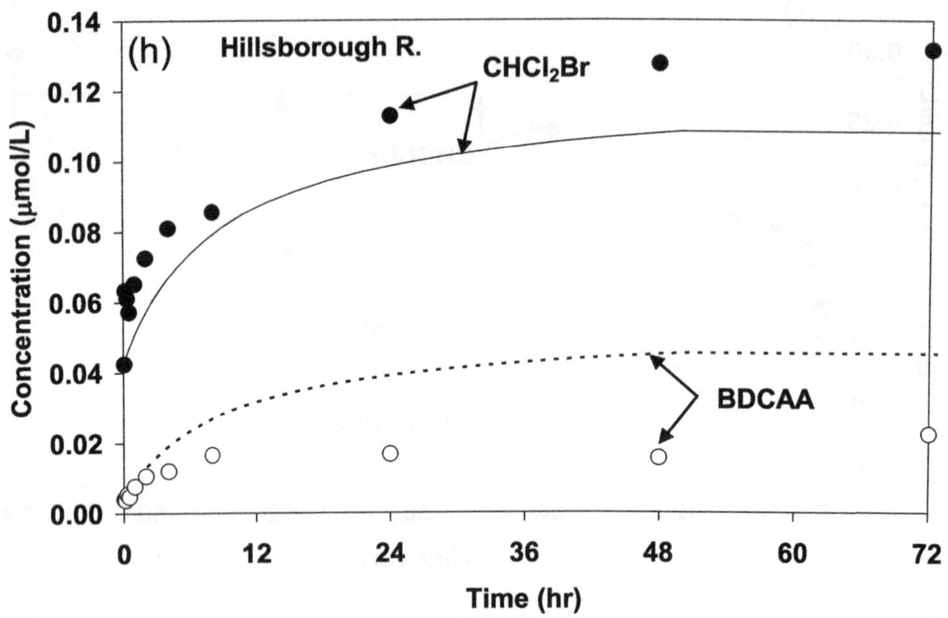

Figure 5.10h Comparison of experimental data (points) and model simulations (lines) for $CHCl_3$, HAAs, and the two brominated DBPs found at the highest concentrations in several chlorinated water sources. (h) $CHCl_2Br$ and BDCAA in Hillsborough River water.

CHAPTER 6
MODELING DBP FORMATION USING DIFFERENTIAL ABSORBANCE AS A MASTER VARIABLE

In this chapter, an alternative approach for estimating DBP formation is presented that complements the predictions of the mechanistic model presented in the preceding chapter. This approach is based on the use of differential absorbance data, which have been shown to correlate very well with DBP concentrations and can therefore be employed to monitor their formation. In the sections that follow, the theoretical fundamentals of the method are presented, and its predictions are compared with experimental data generated for the baseline (LW) water quality.

CONCEPTUAL BASIS FOR THE MODEL

The model of DBP formation considered here (referred to subsequently as the "ΔA model") presumes that DBPs form via the same sequence of reactions considered in the development of the conventional model presented in Chapter 5. However, in addition to the release of $CHCl_3$ and chlorinated HAAs, the ΔA model considers the release of di- and trichloropropanone (DCP and TCP, respectively), dichloroacetonitrile (DCAN), and chloral hydrate (CH) from the chlorinated intermediates, as shown in Figure 6.1. Bromine-containing DBPs are not considered.

In addition to the assumptions used in the conventional modeling (all reactions first-order with respect to the organic species and free chlorine, hence second-order overall), the ΔA model relies on the assumption that the predominant chlorine attack sites in NOM (presumably, aromatic moieties) absorb light at wavelengths < 350 nm. In principle, the model is capable of accounting for absorbance of all the intermediates and the final DBPs as well. However, chemical intuition suggests that, of these species, only the monochlorinated intermediate ICl would be likely to absorb UV light. Furthermore, the modeling results presented below demonstrate that a good fit between the model and the observations can be obtained when the intermediates other than ICl and all the identifiable DBPs are presumed to be non-absorbing at < 350 nm. Therefore, the following discussion is based on the assumption that all the absorbance of the sample derives entirely from the species S and ICl.

The model also presumes that all precursor sites have the same molar absorbance. This assumption can be interpreted as implying that only one type of precursor site exists, or that multiple types of sites exist, but that they all have the same molar absorbance. The fit of the model to the data could undoubtedly be improved by considering two types of sites with different molar absorbances, but as is demonstrated below, the model performs adequately without invoking that complicating factor. A corresponding assumption was made about the absorbance of ICl intermediates.

The presumed reaction pathway leads to a specific relationship between the rate at which reactive sites in NOM molecules are lost (i.e., the rate of decline in the concentration of S) and the rates at which every intermediate and product form as a consequence. In batch or plug flow systems, these rates can be equated with the rate of change of the constituent's concentration. For instance, in a batch system, the following expressions describe the disappearance of S and the appearance and subsequent disappearance of ICl:

$$r_S = \frac{d[S]}{dt} = -k_{ICl}[Cl_{2,\,tot}][S] \tag{6.1}$$

$$r_{ICl} = \frac{d[ICl]}{dt} = (k_{ICl}[S] - (k_{ICl_2} + k_{MCAA})[ICl])[Cl_{2,\,tot}] \tag{6.2}$$

Dividing Equation 6.2 by Equation 6.1 yields:

$$\frac{d[ICl]}{d[S]} = \frac{k_{ICl}[S] - (k_{ICl_2} + k_{MCAA})[ICl]}{[S]} \tag{6.3}$$

Equation 6.3 describes the change in the concentration of ICl as a function of the change in the concentration of S. Significantly, the right-hand side of the equation contains various reaction rate constants and terms for the concentrations of S and ICl, but the concentration of $Cl_{2,tot}$ does not appear as a parameter. As a result, when combined with information about the initial conditions (e.g., the concentrations of S and ICl at $t = 0$), Equation 6.3 can be integrated to characterize the concentration of ICl at any subsequent time, if the concentration of S is known at that time. The time required for those concentrations to be achieved might vary tremendously from one system to the next, depending on the Cl_2 dose, temperature, and other factors, but the relationship between the concentrations of S and ICl would be independent of those factors.

Because all of the reactions in Figure 6.1 are presumed to be first-order in $Cl_{2,tot}$, the general results summarized for ICl in the preceding paragraph apply to all the other intermediates and products in the scheme. That is, in a batch system, assuming the initial conditions are known, the concentration of every intermediate and every DBP can be related to the concentration of unreacted precursor sites that remain. The farther the species of interest is from the precursor sites in the reaction sequence, the more complicated the algebraic result becomes for expressing these relationships, but the general conclusion always applies. Furthermore, by analogy with the results presented by Capellos and Bielski (1972) (who described similar relationships for a complex network of first-order reactions), the equations can be integrated analytically, yielding the long but straightforward expressions shown in Appendix A for the concentrations of all species as a function of $[S_0]$ and $[S]/[S]_0$.

One important feature of the equations shown in Appendix A is that the reaction rate constants never appear in isolation, but only as ratios to the rate constant for disappearance of S (that is, the rate constants always appear in the form k_i/k_{ICl}). This means that, if multiple, kinetically distinct chlorination sites are present, but the ratios of rate constants for sequential and branching reactions are the same for all the sites (as was postulated in our modeling of DBP formation in Chapter 5), the relationships between DBP formation from each type of site and the disappearance of that type of site will be identical for all sites. In that case, the expressions apply to each site individually and also to the sum of all sites (and the total DBP formation).

The expressions relating the concentrations of various DBPs to $[S]_0$ and $[S]/[S]_0$ are conceptually useful in that they suggest a way to relate DBP formation to the disappearance of precursor sites, independent of how long it takes for those sites to disappear and of the chlorine concentration that led to the disappearance. However, unless $[S]_0$ and $[S]/[S]_0$ can be quantified, the derivations are of little practical value. It is in this aspect of the analysis that the change in absorbance of the sample can be used to great advantage, as follows.

Assuming, as indicated previously, that the UV absorbance of the solution includes contributions from only the precursor sites and the monochlorinated intermediate species (ICl), the absolute value of the differential absorbance at any time is given by:

$$|\Delta A_\lambda| = -\Delta A_\lambda = A_{\lambda,0} - A_{\lambda,t} = ([S]_0 - [S]_t)\,\varepsilon_{S,\lambda} - [ICl]_t\,\varepsilon_{ICl,\lambda} \qquad (6.4)$$

where $\varepsilon_{S,\lambda}$ and $\varepsilon_{ICl,\lambda}$ are the molar absorbances of S and ICl, respectively, at wavelength λ.

Once the reaction proceeds to completion and all the precursor sites have reacted, both [S] and [ICl] equal zero, and $|\Delta A_\lambda|$ reaches a maximum value, $|\Delta A_{max,\lambda}|$, i.e.:

$$|\Delta A_{max,\lambda}| = [S]_0\,\varepsilon_{S,\lambda} \qquad (6.5)$$

Combining the above two equations with the equations in Appendix A yields equations for the relationships between the concentrations of the DBPs and $|\Delta A_\lambda|$ that depend only on $[S]_0$, ε_S, ε_{ICl}, and the ratios of the various rate constants to the rate constant for the reaction of the original sites (i.e., k_i/k_{ICl}). In the following section, the DBP data for chlorination of Lake Washington water are used to examine the accuracy of those DBP-$|\Delta A_\lambda|$ relationships.

The fitting was carried out iteratively, using MS Excel® software. Parameters that were varied included the molar absorbance and concentration of precursors, the molar absorbance of the intermediate ICl, and the ratios of the various rate constants. These parameters were altered in ways that achieved a reasonable fit between predictions and observations for the value of $|\Delta A_{max,\,272}|$ (for Lake Washington water, this value was 0.0337 cm^{-1}), the (linear) relationship between $|\Delta A_{max,\,272}|$ and TOX, and the various DBP-$|\Delta A_{272}|$ relationships. Because the fitting was carried out manually, the results are not presumed to represent a global best-fit between the model and experimental observations, but they are suggestive of the quality of the fitting that can be accomplished.

The initial values used for $[S]_0$ and the k_i/k_{ICl} ratios were based on those presented in Chapter 5. However, in that case, no values of $\varepsilon_{S,272}$ and $\varepsilon_{ICl,272}$ provided satisfactory fits to the DBP vs. $|\Delta A_{272}|$ data. Therefore, the fitting parameters were adjusted without any constraints, in an effort to match the experimental observations. The values that were ultimately identified as providing a good match are presented in Table 6.1. The value of 3.66×10^{-6} M for the concentration of precursor sites in Lake Washington is 22% larger than the value used in the conventional modeling (Chapter 5). Since the molecular weight of NOM molecules is not known, the molar absorbance at 272 nm shown in the table is normalized to the number of sites, i.e., the absorbance is 9300 cm^{-1} per mol/L of sites. Assuming, as a rough estimate, that the molecules contain one site for each aromatic ring, this value of the molar absorbance is close to that of highly activated aromatic groups in pure compounds (Li, Benjamin, and Korshin 2000). Thus, the estimates of $[S]_0$, $\varepsilon_{S,272}$, and $\varepsilon_{ICl,272}$ are all reasonable and consistent with estimates that might be made using other approaches. On the other hand, the ratios of the reaction rate constants (k_i/k_{ICl}) are, in some cases, very different from the values used in the conventional kinetic modeling; no basis exists for deciding which, if either, of these sets of values is realistic.

Prior research has indicated that the concentration of TOX generated when water is chlorinated is highly correlated with $|\Delta A_{272}|$ (Korshin, Li, and Benjamin 1997a). The relationship is characterized by a linear function passing through the origin, with a slope typically close to 3.1×10^{-4} mol TOX/cm^{-1}, independent of water quality. As shown in Figure 6.2, the TOX vs. $|\Delta A_{272}|$ correlations generated by the model were quite sensitive to the molar absorbance of the intermediate

Table 6.1
Parameters used in the model calculations

Parameter	Value	Parameter	Value
$[S]_0$	3.66×10^{-6} M	k_{DCAA}/k_{ICl}	2.0
$\varepsilon_{S,272}$	9300 $M^{-1}cm^{-1}$	k_{DCAN}/k_{ICl}	0.3
$\varepsilon_{ICl,272}$	7300 $M^{-1}cm^{-1}$	k_{DCP}/k_{ICl}	0.2
k_{ICl_2}/k_{ICl}	3.3	k_{CHCl_3}/k_{ICl}	5.3
k_{ICl_3}/k_{ICl}	12.7	k_{TCAA}/k_{ICl}	5.0
k_{PCl_3}/k_{ICl}	12.7	k_{CH}/k_{ICl}	0.7
k_{MCAA}/k_{ICl}	0.1	k_{TCP}/k_{ICl}	0.3

ICl, especially at low values of $|\Delta A_{272}|$. Assigning $\varepsilon_{ICl,272}$ a value of 7300 $M^{-1}cm^{-1}$ led to almost perfect agreement between the model and experimental TOX-$|\Delta A_{272}|$relationship (Figure 6.3).

The experimental data and model predictions for the DBP vs. $|\Delta A_{272}|$ relationships are compared in Figures 6.4 to 6.11. The model reproduces the experimental observations for virtually all the identified, chlorinated DBPs (with the exception of those susceptible to rapid spontaneous decay, such as DCP). Note that the data in the figures characterize chlorination under widely varying halogenation conditions, including reaction times from 5 minutes to 7 days, and Cl_2:DOC ratios from 0.1 to 3.0 mg/mg.

In addition to the usefulness of the model for understanding DBP formation in those cases where the model reproduces the experimental results well, the situations in which the model and experimental results do not match can also provide important information. For instance, the accumulation and subsequent disappearance of DCP, TCP, and DCAN as $|\Delta A_{272}|$ increases is a clear indication of the formation and decay of those DBPs during the course of the reaction. The value of $|\Delta A_{272}|$ at which the decay begins might provide clues as to the conditions necessary for those decay reactions to proceed. In addition, by adding a decay reaction for these species to the overall scheme, it is likely that the model could be used to predict their concentrations throughout the reaction period. Similarly, Figure 6.11 shows that the model fit to the chloral hydrate data is not as good as it is for most other stable DBPs. This result suggests that CH is not released by cleavage of ICl_3, and that a more complex reaction sequence may be involved in formation of that compound. On the other hand, the success of the modeling effort with respect to formation of THMs and HAAs indicates that the four-step sequence employed by the model is adequate for characterizing the formation of the DBPs of greatest interest.

While the ΔA model is not a substitute for the traditional kinetic description of DBP formation, important conclusions regarding the kinetics of NOM halogenation can nonetheless be drawn based on the model. For instance, the k_i/k_{ICl} ratios shown in Table 6.1 indicate that the first chlorine incorporation step (formation of ICl) is the rate-controlling step for formation of all the DBPs, in accord with the conventional kinetic modeling. Also, at least in this preliminary modeling effort, the rate constants for halogenation increase with the number of chlorine atoms incorporated into the intermediates.

One reason that the values of the k_i/k_{ICl} ratios shown in Table 6.1 do not conform to those generated in the kinetic modeling is that the values in the ΔA model, in effect, represent averaged rate constants for all types of reaction sites in the NOM molecules, whereas the conventional model considered two kinetically distinct sites. Separation of the contributions of those two types of sites and considering them to have different molar absorbances might improve the coherence of the two modeling approaches. Efforts to improve that coherence should be a focus of future effort.

The above discussion addresses only the formation of chlorinated DBPs. However, as was demonstrated by several figures in Chapter 4, DBP vs. $|\Delta A_{272}|$ correlations have similar strength and functional form for brominated THMs and HAAs. Although the ΔA model has not yet been applied quantitatively to brominated species, further development of the model certainly could extend it to those and other DBPs.

Finally, it is worth noting that the ΔA model could be relatively easily converted into a conventional kinetic model, and vice versa. The former transformation could be accomplished simply by assigning a numerical value to k_{ICl} and carrying out the relevant calculations. Thus, the two types of model are not mutually exclusive or competitive with one another, but rather complementary; in effect, they provide two different and useful views of the same process.

SUMMARY

In this chapter, a model has been developed to explain the observed correlations between the formation of chlorine-containing DBPs and the differential absorbance associated with the chlorination process. The experimental data for most of the DBP-$|\Delta A_{272}|$ relationships could be fit very well by assuming that the unaltered sites and the first intermediate species (presumably a mono-chlorinated, aromatic compound) absorb UV light, but that intermediates and products formed by subsequent chlorination reactions do not. The estimated molar absorbance of the unaltered, reactive sites in NOM is in the same range as that of highly activated pure aromatic compounds, with the chlorinated intermediate being somewhat less UV-absorbing than the unaltered sites.

Consistent with the conventional kinetics analysis, the ΔA analysis indicates that formation of the first halogenated intermediate is the rate-limiting step for DBP formation. It also suggests that the breakdown of the mono-, di- and trichlorinated intermediates becomes progressively easier in that order.

The fit of the model to the data for formation of chloral hydrate (CH) was less good than that for other stable DBPs, suggesting that formation of CH does not form from the same immediate precursor as $CHCl_3$ and TCAA. Additional modeling of this relationship is likely to provide insight into the pathway by which CH forms. The other DBPs for which the model did not reproduce the experimental data were those which clearly underwent decay reactions after they formed. Further development of the model would likely lead to insights into those processes as well.

Aside from the potential value of the ΔA model as a tool for understanding the chemistry of DBP formation, the practical value of the model is associated with the technical simplicity of absorbance measurements compared to conventional chromatographic methods. Because of this simplicity, the ΔA model allows empirical estimates of DBP formation to be made rapidly and inexpensively, thereby greatly diminishing the need for sophisticated, predictive kinetic models. Furthermore, careful analysis of the ΔA vs. time relationship may well facilitate the development of better, formal kinetic models of DBP formation in the future. Additional developmental work is needed to extend the ΔA model to the formation of brominated DBPs and to DBPs that undergo decay reactions after they form, and also to improve the coherence between the ΔA model and conventional kinetic modeling.

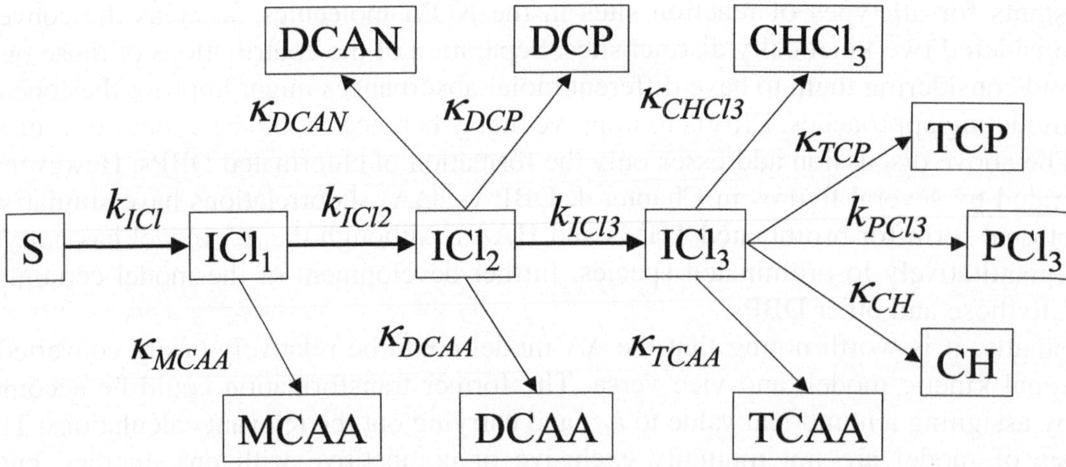

Figure 6.1 The proposed four-step chlorination of an NOM site, showing the formation of chlorinated intermediates and reaction end products

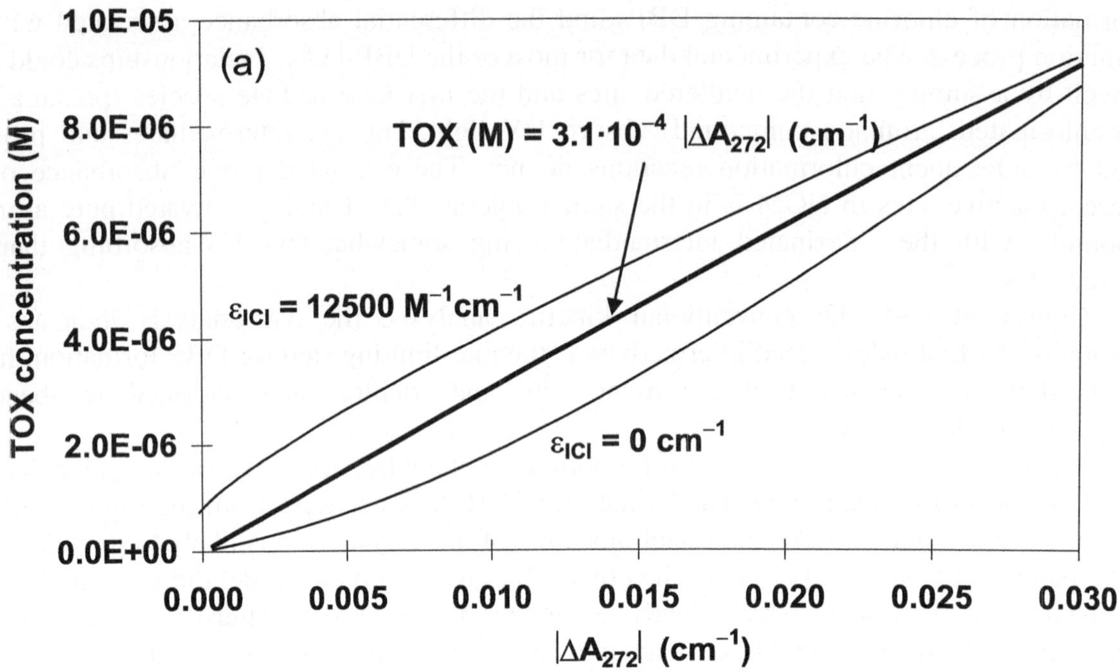

Figure 6.2a Influence of the molar absorbance of intermediate ICl on the model prediction for the relationship between TOX and $|\Delta A_{272}|$. (a) Values of $|\Delta A_{272}|$ up to 0.03 cm^{-1}. Molar absorbance of the unaltered site is 9300 $M^{-1}cm^{-1}$.

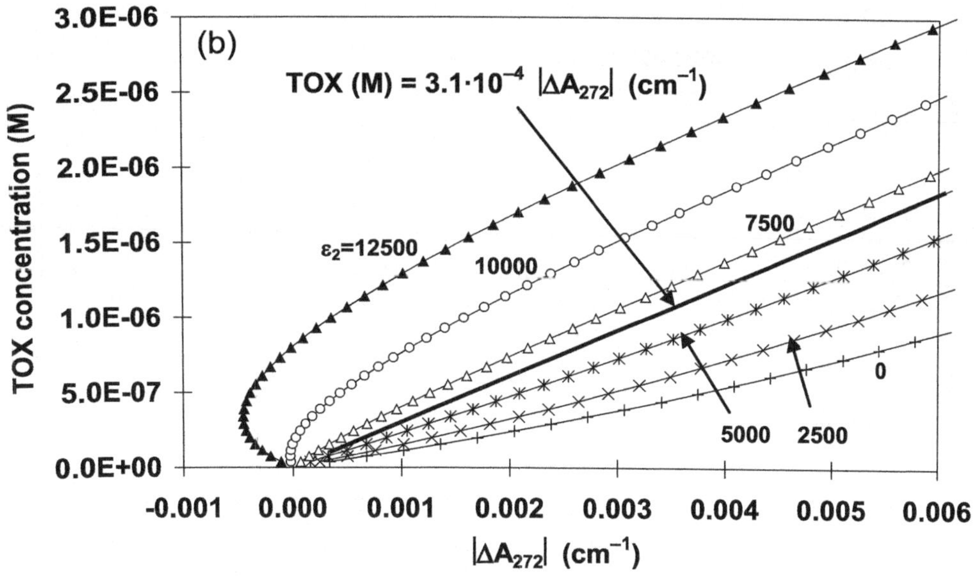

Figure 6.2b Influence of the molar absorbance of intermediate ICl on the model prediction for the relationship between TOX and $|\Delta A_{272}|$. (b) Low values of $|\Delta A_{272}|$ (up to 0.006 cm^{-1}). Molar absorbance of the unaltered site is 9300 $M^{-1}cm^{-1}$.

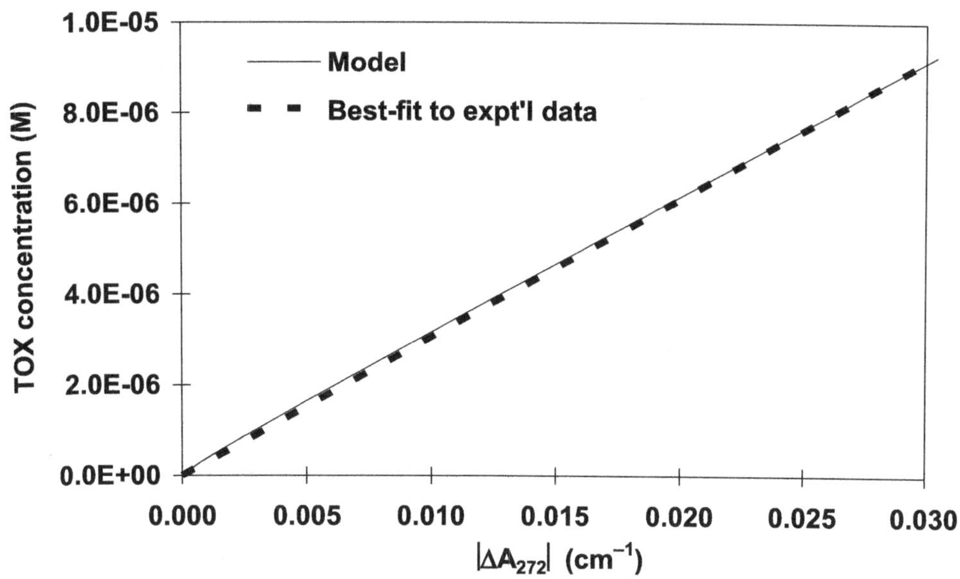

Figure 6.3 Comparison of model calculations for TOX-$|\Delta A_{272}|$ relationship with typical experimental measurements. Molar absorbance of the unaltered site and ICl for the model calculations are 9300 and 7300 $M^{-1}cm^{-1}$, respectively.

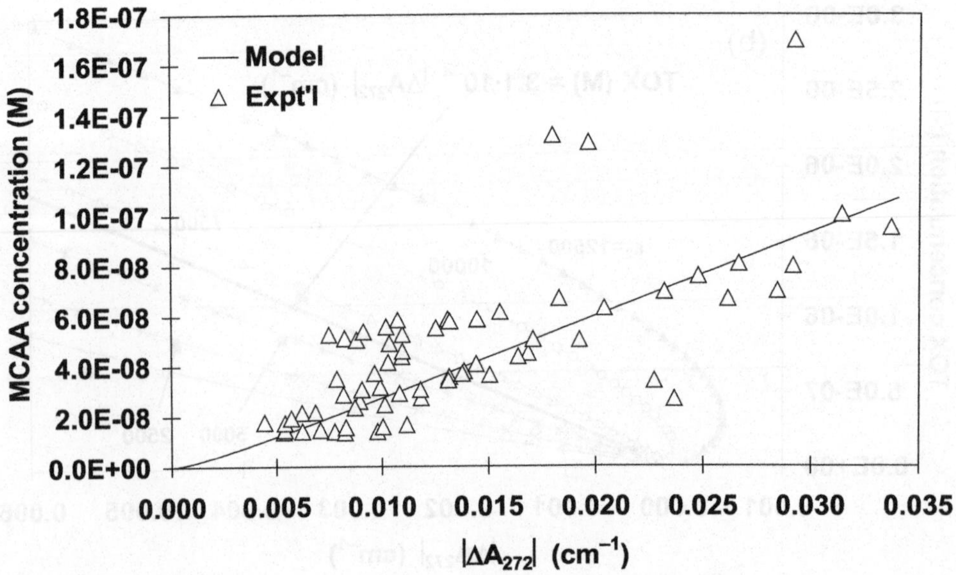

Figure 6.4 Comparison of model calculations for MCAA-$|\Delta A_{272}|$ relationship with experimental measurements

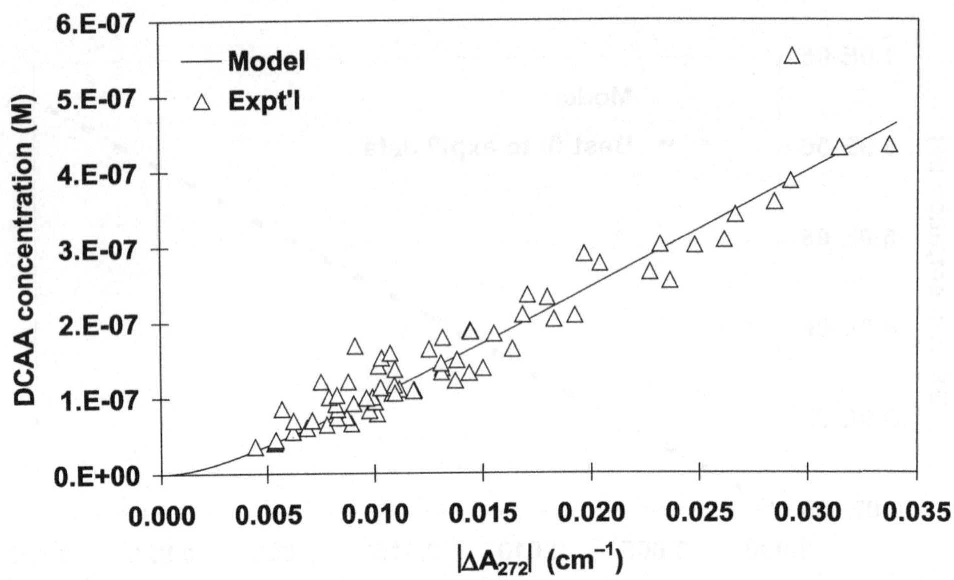

Figure 6.5 Comparison of model calculations for DCAA-$|\Delta A_{272}|$ relationship with experimental measurements

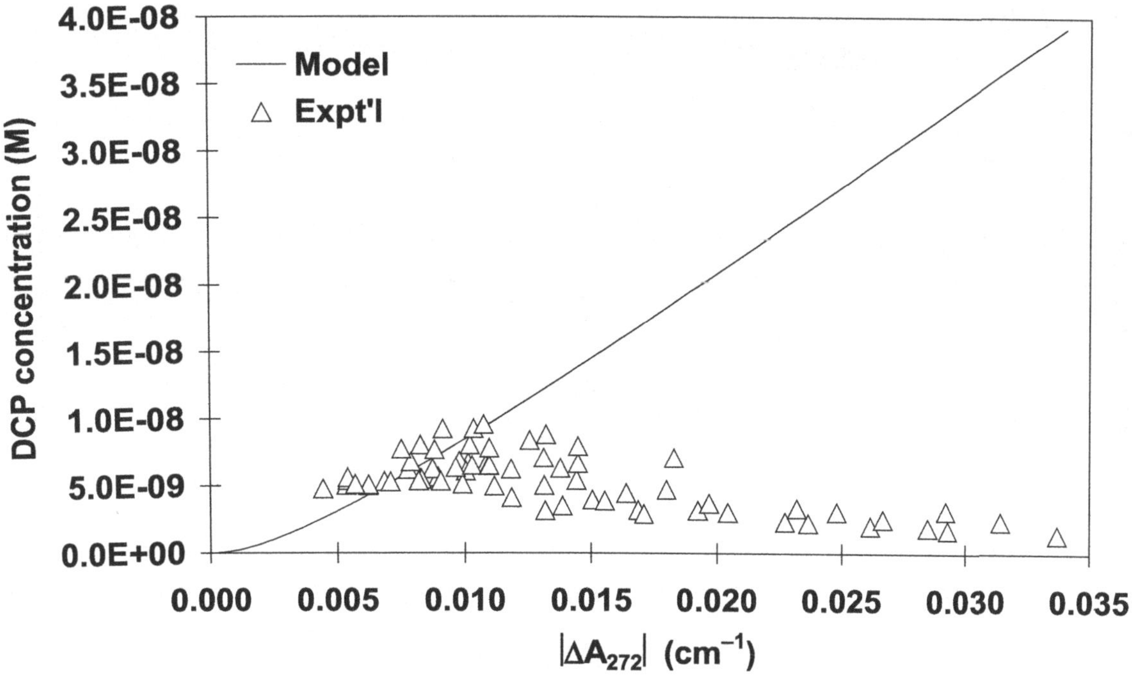

Figure 6.6 Comparison of model calculations for DCP-$|\Delta A_{272}|$ relationship with experimental measurements

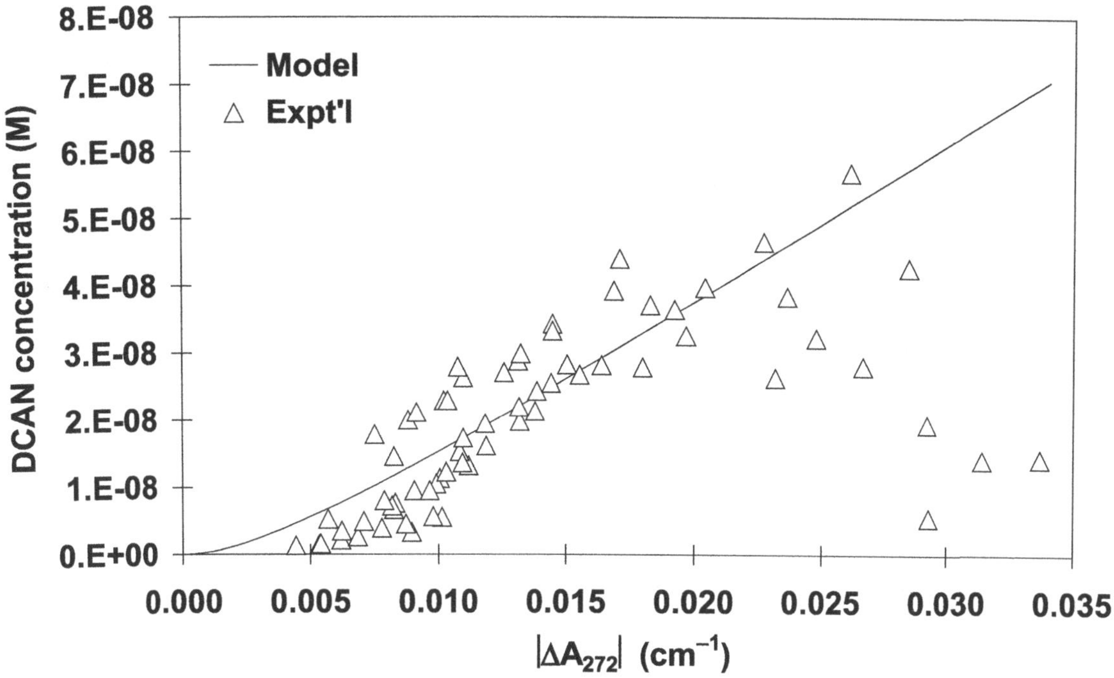

Figure 6.7 Comparison of model calculations for DCAN-$|\Delta A_{272}|$ relationship with experimental measurements

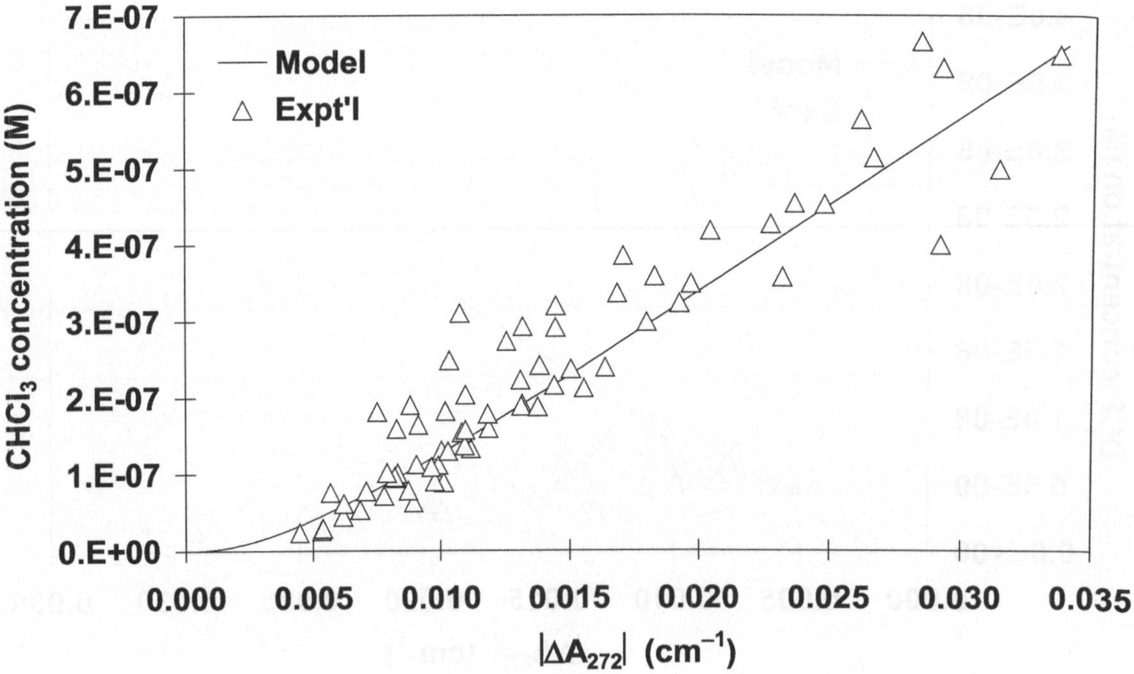

Figure 6.8 Comparison of model calculations for $CHCl_3$-$|\Delta A_{272}|$ relationship with experimental measurements

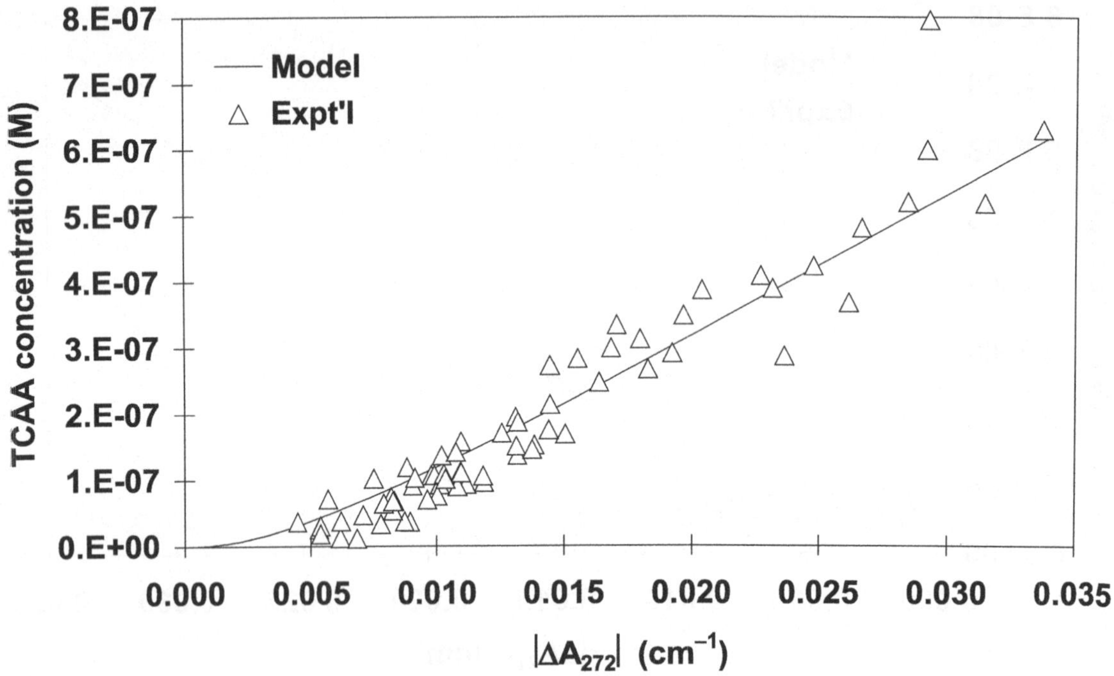

Figure 6.9 Comparison of model calculations for TCAA-$|\Delta A_{272}|$ relationship with experimental measurements

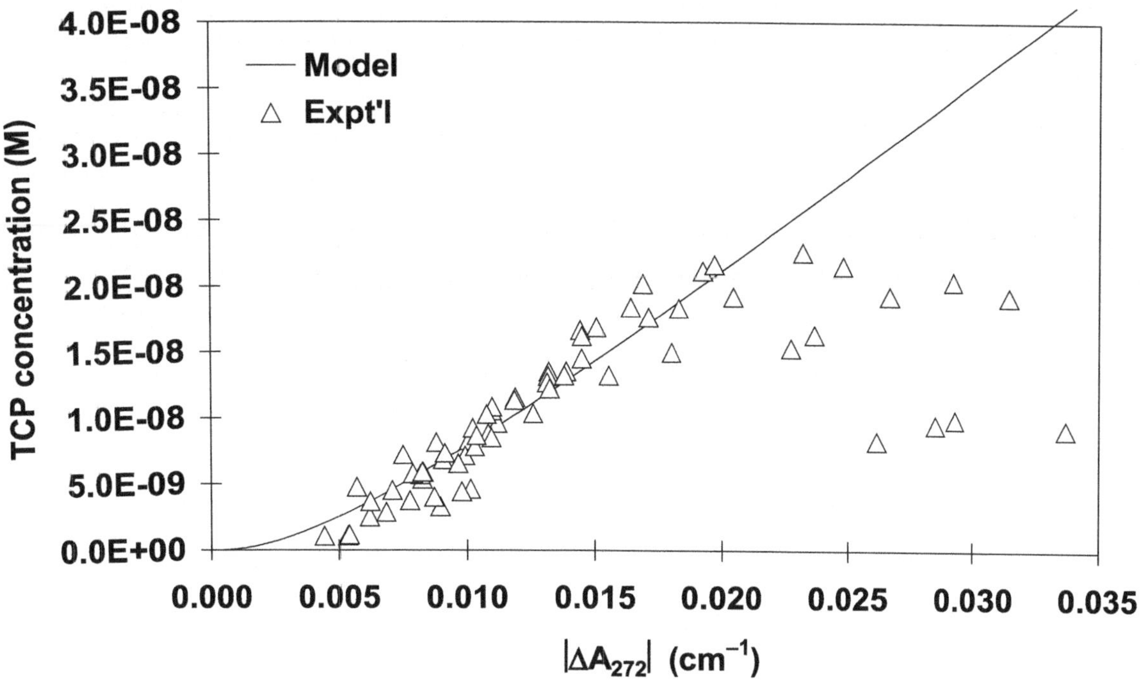

Figure 6.10 Comparison of model calculations for TCP-$|\Delta A_{272}|$ relationship with experimental measurements

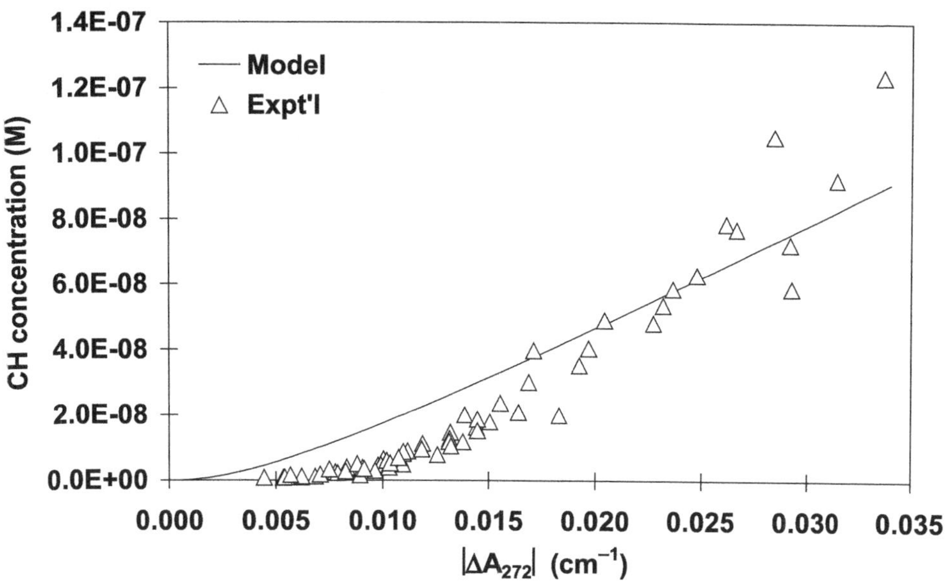

Figure 6.11 Comparison of model calculations for CH-$|\Delta A_{272}|$ relationship with experimental measurements

CHAPTER 7
SUMMARY, CONCLUSIONS AND RECOMMENDATIONS

This research project included experimental and theoretical studies addressing several issues related to DBP formation in drinking water. The primary accomplishment of the project was the formulation and testing of a scientifically sound, mechanistic model for the reactions of DBP precursor sites with halogen species. The central reaction pathway in the model is a four-step sequence, consisting of three steps in which halogen atoms are incorporated into a reactive site in NOM, and a fourth step in which a portion of the molecule that includes the trihalogenated site is cleaved to release a THM, trihalogenated HAA, or a different, unidentified trihalogenated DBP. In addition, branching reactions are envisioned to allow the intermediates along the main pathway to release mono- or di-halogenated HAAs. If the raw water contains bromide ions, the model is still applicable, but it includes several parallel pathways, differing with respect to whether bromine or chlorine is incorporated in each step.

A wide range of reaction conditions for the formation of DBPs was explored experimentally (varying reaction times, pH, halogen dose, temperature), and the ability of the model to reproduce the experimental results was evaluated. When the assumption was made that two kinetically distinct types of precursor sites exist in NOM, each of which reacts according to the pathway described above, the model was able to reproduce the kinetics of DBP formation over time frames from minutes to days, as long as the Cl_2 dose was sufficient to maintain a free Cl_2 residual throughout the time frame of interest and the Br^- concentration of the raw water was low. If either of these latter restrictions was violated, then the model could still reproduce the experimental data fairly well, but only if a third type of site was postulated to account for a very rapid component of DBP formation. The model was also shown to be applicable for characterizing DBP formation in a variety of dissimilar water qualities.

Based on the tests conducted in this research, the "fast" reaction sites constitute ca. 30% of the precursor sites in NOM. Under typical chlorination conditions, the reactions of the fast sites are almost complete in less than one hour, whereas the slow sites continue to react over a period of a few days. The reactions of the very rapidly reacting sites appear to be complete so rapidly that they can be considered to be instantaneous; also, the concentration of these sites is very small.

It was possible to model the formation kinetics of THMs and HAAs containing bromine using a single, constant value for the ratio of the rates of bromination and chlorination (k_{Br} / k_{Cl}) for all reactions other than the final, DBP-formation step. The value of this so-called branching ratio was typically > 10, which attests to a significantly greater reactivity of bromine species than chlorine species in halogenation reactions, as has been suggested by prior researchers. The ratio k_{Br} / k_{Cl} appeared to be even larger for the final, DBP-formation step.

Overall, the model developed in this study utilizes 13 independent parameters to model the formation of DBPs under a wide range of reaction conditions. These parameters include the total concentration of all reaction sites, the distribution of those sites between fast- and slow- reacting sites, the ratio of kinetic constants of the two types of sites, two parameters characterizing the relative reactivity of bromine species to chlorine species, and eight kinetic rate constants.

A unique aspect of the current project was the characterization of NOM reactions with free chlorine over extremely short periods (< 1 s), using quench-flow and stopped-flow technology. These experiments indicated that the spectroscopic properties of the fast, slow and even the instantaneously reacting sites are remarkably uniform. In other words, the kinetic dissimilarities of the various halogen attack sites in NOM does not appear to reflect comparably large differences in their chemical nature.

In addition to predicting DBP concentrations, an attempt was made to use the model to reproduce the strong relationship between DBP formation and the change in the UV absorbance (i.e., the differential absorbance, ΔA) of water samples when they are chlorinated. Only the data for chlorination of Lake Washington water, which contains very low bromide concentrations, were used in this part of the study. It was possible to reproduce most of the DBP-$|\Delta A_{272}|$ relationships quite well within the same context as the kinetic model (i.e., the four-step sequence, with branching). This success was achieved not only for the relationships of $|\Delta A_{272}|$ with THMs and HAAs, but also for those with several trace DBPs.

The modeling effort for the absorbance data suggested that the molar absorbance of the unaltered precursor sites is comparable to that of highly activated pure compounds, and that approximately 20% and 80% of the absorbance is destroyed in the first and second halogen incorporation steps, respectively. This modeling also suggested that the formation pathway for chloral hydrate is slightly different from that for other trihalogenated DBPs, perhaps involving an additional intermediate step. The values of the reaction constants needed to fit the absorbance data were, in some cases, substantially different from those needed to fit the concentration data. Additional research is required to resolve this issue.

The models developed in this research project are, of course, not free of shortcomings and/or deficiencies. The main shortcomings are the need to use a fitting function to predict chlorine decay and to evaluate the fitting parameters for a given water quality. However, the available data indicate that the fitting function for chlorine decay is smooth and rather easily evaluated, and that the values of the rate constants fall within reasonably narrow windows. Evaluation of these parameters in a wider range of waters will undoubtedly provide guidance for predicting their approximate values without carrying out a full-blown experimental program at each application site.

Another deficiency is the inability to model DBP formation kinetics and the relationships between DBP formation and changes in absorbance using a consistent parameter set. While each of these types of correlations can be used independently and successfully, the disconnect between the values of the parameters used in the two models is a clear indication of some fundamental flaw in the model construct and is an area ripe for further investigation.

Future research should focus on the further development of the mechanistic approach, further elucidation of the effects of pH and temperature on the values of the rate constants, and enhancement of the conceptual basis of the model so that it can explain both kinetic and spectroscopic observations. It is also necessary to apply the mechanistic model to a wider variety of water qualities, especially treated waters, so that the effects of water treatment can be ascertained. The integration of such data for a representative range of water qualities and reaction conditions with the mechanistic model will constitute an important advance in the evolution of drinking water chemistry and its engineering applications.

APPENDIX A
EXPRESSIONS FOR THE CONCENTRATIONS OF INTERMEDIATES AND FINAL PRODUCTS OF DBP-FORMING REACTIONS IN TERMS OF PRESCURSOR SITE CONCENTRATIONS AND RELATIVE REACTION RATE CONSTANTS

Definition of symbols in the following equations:

S_0 and S: concentration of the chlorine attack sites present initially (S_0) and at any subsequent time (S)

ICl, ICl_2 and ICl_3: concentraion of mono-, di- and tri-chlorinated intermediates

PCl_3: an unidentified tri-chlorinated product formed from ICl_3

k_i: reaction rate constant or formation of species i

K_i: sum of all reaction rate constants associated with the beakdown of species ICl_{i-1}; e.g., K_3 is the sum of the rate constants for breakdown of ICl_2

$$[\mathrm{ICl}] = k_{\mathrm{ICl}}[\mathrm{S}]_0\left[\frac{1}{(K_2-k_{\mathrm{ICl}})}\left(\frac{[\mathrm{S}]}{[\mathrm{S}]_0}\right)+\frac{1}{(k_{\mathrm{ICl}}-K_2)}\left(\frac{[\mathrm{S}]}{[\mathrm{S}]_0}\right)^{\frac{K_2}{k_{\mathrm{ICl}}}}\right] \tag{A.1}$$

$$[\mathrm{MCAA}] = \gamma_{MCAA}k_{\mathrm{ICl}}[\mathrm{S}]_0\left[\frac{1}{k_{\mathrm{ICl}}K_2}-\left(\frac{\left(\frac{[\mathrm{S}]}{[\mathrm{S}]_0}\right)}{k_{\mathrm{ICl}}(k_{\mathrm{ICl}}-K_2)}-\frac{\left(\frac{[\mathrm{S}]}{[\mathrm{S}]_0}\right)^{\frac{K_2}{k_{\mathrm{ICl}}}}}{K_2(K_2-k_{\mathrm{ICl}})}\right)\right] \tag{A.2}$$

$$[\mathrm{ICl}_2] = k_{\mathrm{ICl}}k_{\mathrm{ICl}_2}[\mathrm{S}]_0\left[\frac{\left(\frac{[\mathrm{S}]}{[\mathrm{S}]_0}\right)}{(K_2-k_{\mathrm{ICl}})(K_3-k_{\mathrm{ICl}})}+\frac{\left(\frac{[\mathrm{S}]}{[\mathrm{S}]_0}\right)^{\frac{K_2}{k_{\mathrm{ICl}}}}}{(k_{\mathrm{ICl}}-K_2)(K_3-K_2)}+\frac{\left(\frac{[\mathrm{S}]}{[\mathrm{S}]_0}\right)^{\frac{K_3}{k_{\mathrm{ICl}}}}}{(k_{\mathrm{ICl}}-K_3)(K_2-K_3)}\right] \tag{A.3}$$

$$[\mathrm{DCAA}] = \gamma_{\mathrm{DCAN}} k_{\mathrm{ICl}} k_{\mathrm{ICl_2}} [\mathrm{S}]_0 \left[\frac{1}{k_{\mathrm{ICl}}(K_2 - k_{\mathrm{ICl}})(K_3 - k_{\mathrm{ICl}})} + \frac{1}{K_2(k_{\mathrm{ICl}} - K_2)(K_3 - K_2)} + \frac{1}{K_3(k_{\mathrm{ICl}} - K_3)(K_2 - K_3)} - \frac{\left(\frac{[\mathrm{S}]}{[\mathrm{S}]_0}\right)}{k_{\mathrm{ICl}}(K_2 - k_{\mathrm{ICl}})(K_3 - k_{\mathrm{ICl}})} - \frac{\left(\frac{[\mathrm{S}]}{[\mathrm{S}]_0}\right)^{\frac{K_2}{k_{\mathrm{ICl}}}}}{K_2(k_{\mathrm{ICl}} - K_2)(K_3 - K_2)} - \frac{\left(\frac{[\mathrm{S}]}{[\mathrm{S}]_0}\right)^{\frac{K_3}{k_{\mathrm{ICl}}}}}{K_3(k_{\mathrm{ICl}} - K_3)(K_2 - K_3)} \right] \quad (A.4)$$

$$[\mathrm{DCAN}] = \gamma_{\mathrm{DCAN}} k_{\mathrm{ICl}} k_{\mathrm{ICl_2}} [\mathrm{S}]_0 \left[\frac{1}{k_{\mathrm{ICl}}(K_2 - k_{\mathrm{ICl}})(K_3 - k_{\mathrm{ICl}})} + \frac{1}{K_2(k_{\mathrm{ICl}} - K_2)(K_3 - K_2)} + \frac{1}{K_3(k_{\mathrm{ICl}} - K_3)(K_2 - K_3)} - \frac{\left(\frac{[\mathrm{S}]}{[\mathrm{S}]_0}\right)}{k_{\mathrm{ICl}}(K_2 - k_{\mathrm{ICl}})(K_3 - k_{\mathrm{ICl}})} - \frac{\left(\frac{[\mathrm{S}]}{[\mathrm{S}]_0}\right)^{\frac{K_2}{k_{\mathrm{ICl}}}}}{K_2(k_{\mathrm{ICl}} - K_2)(K_3 - K_2)} - \frac{\left(\frac{[\mathrm{S}]}{[\mathrm{S}]_0}\right)^{\frac{K_3}{k_{\mathrm{ICl}}}}}{K_3(k_{\mathrm{ICl}} - K_3)(K_2 - K_3)} \right] \quad (A.5)$$

$$[\mathrm{DCP}] = \gamma_{\mathrm{DCP}} k_{\mathrm{ICl}} k_{\mathrm{ICl_2}} [\mathrm{S}]_0 \left[\frac{1}{k_{\mathrm{ICl}}(K_2 - k_{\mathrm{ICl}})(K_3 - k_{\mathrm{ICl}})} + \frac{1}{K_2(k_{\mathrm{ICl}} - K_2)(K_3 - K_2)} + \frac{1}{K_3(k_{\mathrm{ICl}} - K_3)(K_2 - K_3)} \right. \\ \left. - \frac{\left(\frac{[\mathrm{S}]}{[\mathrm{S}]_0}\right)}{k_{\mathrm{ICl}}(K_2 - k_{\mathrm{ICl}})(K_3 - k_{\mathrm{ICl}})} - \frac{\left(\frac{[\mathrm{S}]}{[\mathrm{S}]_0}\right)^{\frac{K_2}{k_{\mathrm{ICl}}}}}{K_2(k_{\mathrm{ICl}} - K_2)(K_3 - K_2)} - \frac{\left(\frac{[\mathrm{S}]}{[\mathrm{S}]_0}\right)^{\frac{K_3}{k_{\mathrm{ICl}}}}}{K_3(k_{\mathrm{ICl}} - K_3)(K_2 - K_3)} \right] \tag{A.6}$$

$$[\mathrm{PCl_3}] = k_{\mathrm{PCl_3}} k_{\mathrm{ICl}} k_{\mathrm{ICl_2}} k_{\mathrm{ICl_3}} [\mathrm{S}]_0 \left[\frac{1}{k_{\mathrm{ICl}} K_2 K_3 K_4} - \frac{\left(\frac{[\mathrm{S}]}{[\mathrm{S}]_0}\right)}{k_{\mathrm{ICl}}(K_2 - k_{\mathrm{ICl}})(K_3 - k_{\mathrm{ICl}})(K_4 - k_{\mathrm{ICl}})} - \frac{\left(\frac{[\mathrm{S}]}{[\mathrm{S}]_0}\right)^{\frac{K_2}{k_{\mathrm{ICl}}}}}{K_2(k_{\mathrm{ICl}} - K_2)(K_3 - K_2)(K_4 - K_2)} \right. \\ \left. - \frac{\left(\frac{[\mathrm{S}]}{[\mathrm{S}]_0}\right)^{\frac{K_3}{k_{\mathrm{ICl}}}}}{K_3(k_{\mathrm{ICl}} - K_3)(K_2 - K_3)(K_4 - K_3)} - \frac{\left(\frac{[\mathrm{S}]}{[\mathrm{S}]_0}\right)^{\frac{K_4}{k_{\mathrm{ICl}}}}}{K_4(k_{\mathrm{ICl}} - K_4)(K_2 - K_4)(K_3 - K_4)} \right] \tag{A.7}$$

$$[\mathrm{CHCl_3}] = k_{\mathrm{CHCl_3}} k_{\mathrm{ICl}} k_{\mathrm{ICl_2}} k_{\mathrm{ICl_3}} [\mathrm{S}]_0 \left[\frac{1}{k_{\mathrm{ICl}} K_2 K_3 K_4} - \frac{\left(\frac{[\mathrm{S}]}{[\mathrm{S}]_0}\right)}{k_{\mathrm{ICl}}(K_2 - k_{\mathrm{ICl}})(K_3 - k_{\mathrm{ICl}})(K_4 - k_{\mathrm{ICl}})} - \frac{\left(\frac{[\mathrm{S}]}{[\mathrm{S}]_0}\right)^{\frac{K_2}{k_{\mathrm{ICl}}}}}{K_2(k_{\mathrm{ICl}} - K_2)(K_3 - K_2)(K_4 - K_2)} \right. \left. - \frac{\left(\frac{[\mathrm{S}]}{[\mathrm{S}]_0}\right)^{\frac{K_3}{k_{\mathrm{ICl}}}}}{K_3(k_{\mathrm{ICl}} - K_3)(K_2 - K_3)(K_4 - K_3)} - \frac{\left(\frac{[\mathrm{S}]}{[\mathrm{S}]_0}\right)^{\frac{K_4}{k_{\mathrm{ICl}}}}}{K_4(k_{\mathrm{ICl}} - K_4)(K_2 - K_4)(K_3 - K_4)} \right] \quad (\mathrm{A.8})$$

$$[\mathrm{TCAA}] = k_{\mathrm{TCAA}} k_{\mathrm{ICl}} k_{\mathrm{ICl_2}} k_{\mathrm{ICl_3}} [\mathrm{S}]_0 \left[\frac{1}{k_{\mathrm{ICl}} K_2 K_3 K_4} - \frac{\left(\frac{[\mathrm{S}]}{[\mathrm{S}]_0}\right)}{k_{\mathrm{ICl}}(K_2 - k_{\mathrm{ICl}})(K_3 - k_{\mathrm{ICl}})(K_4 - k_{\mathrm{ICl}})} - \frac{\left(\frac{[\mathrm{S}]}{[\mathrm{S}]_0}\right)^{\frac{K_2}{k_{\mathrm{ICl}}}}}{K_2(k_{\mathrm{ICl}} - K_2)(K_3 - K_2)(K_4 - K_2)} \right. \left. - \frac{\left(\frac{[\mathrm{S}]}{[\mathrm{S}]_0}\right)^{\frac{K_3}{k_{\mathrm{ICl}}}}}{K_3(k_{\mathrm{ICl}} - K_3)(K_2 - K_3)(K_4 - K_3)} - \frac{\left(\frac{[\mathrm{S}]}{[\mathrm{S}]_0}\right)^{\frac{K_4}{k_{\mathrm{ICl}}}}}{K_4(k_{\mathrm{ICl}} - K_4)(K_2 - K_4)(K_3 - K_4)} \right] \quad (\mathrm{A.9})$$

$$[\mathrm{CH}] = k_{\mathrm{CH}} k_{\mathrm{ICl}} k_{\mathrm{ICl_2}} k_{\mathrm{ICl_3}} [\mathrm{S}]_0 \left[\frac{1}{k_{\mathrm{ICl}} K_2 K_3 K_4} - \frac{\left(\frac{[\mathrm{S}]}{[\mathrm{S}]_0}\right)}{k_{\mathrm{ICl}}(K_2 - k_{\mathrm{ICl}})(K_3 - k_{\mathrm{ICl}})(K_4 - k_{\mathrm{ICl}})} - \frac{\left(\frac{[\mathrm{S}]}{[\mathrm{S}]_0}\right)^{\frac{K_2}{k_{\mathrm{ICl}}}}}{K_2(k_{\mathrm{ICl}} - K_2)(K_3 - K_2)(K_4 - K_2)} - \frac{\left(\frac{[\mathrm{S}]}{[\mathrm{S}]_0}\right)^{\frac{K_3}{k_{\mathrm{ICl}}}}}{K_3(k_{\mathrm{ICl}} - K_3)(K_2 - K_3)(K_4 - K_3)} - \frac{\left(\frac{[\mathrm{S}]}{[\mathrm{S}]_0}\right)^{\frac{K_4}{k_{\mathrm{ICl}}}}}{K_4(k_{\mathrm{ICl}} - K_4)(K_2 - K_4)(K_3 - K_4)} \right] \quad (A.10)$$

$$[\mathrm{TCP}] = k_{\mathrm{TCP}} k_{\mathrm{ICl}} k_{\mathrm{ICl_2}} k_{\mathrm{ICl_3}} [\mathrm{S}]_0 \left[\frac{1}{k_{\mathrm{ICl}} K_2 K_3 K_4} - \frac{\left(\frac{[\mathrm{S}]}{[\mathrm{S}]_0}\right)}{k_{\mathrm{ICl}}(K_2 - k_{\mathrm{ICl}})(K_3 - k_{\mathrm{ICl}})(K_4 - k_{\mathrm{ICl}})} - \frac{\left(\frac{[\mathrm{S}]}{[\mathrm{S}]_0}\right)^{\frac{K_2}{k_{\mathrm{ICl}}}}}{K_2(k_{\mathrm{ICl}} - K_2)(K_3 - K_2)(K_4 - K_2)} - \frac{\left(\frac{[\mathrm{S}]}{[\mathrm{S}]_0}\right)^{\frac{K_3}{k_{\mathrm{ICl}}}}}{K_3(k_{\mathrm{ICl}} - K_3)(K_2 - K_3)(K_4 - K_3)} - \frac{\left(\frac{[\mathrm{S}]}{[\mathrm{S}]_0}\right)^{\frac{K_4}{k_{\mathrm{ICl}}}}}{K_4(k_{\mathrm{ICl}} - K_4)(K_2 - K_4)(K_3 - K_4)} \right] \quad (A.11)$$

where:

$$K_2 = k_{ICl_2} + k_{MCAA} \quad \text{(A.12)}$$

$$K_3 = k_{ICl_3} + k_{DCAA} + k_{DCAN} + k_{DCP} \quad \text{(A.13)}$$

$$K_4 = k_{PCl_3} + k_{TCAA} + k_{TCP} + k_{CHCl_3} + k_{CH} \quad \text{(A.14)}$$

REFERENCES

Adin, A., J. Katzhendler, D. Alkaslassy, and C. Rav-Acha. 1991. THM Formation in Chlorinated Drinking Water: A Kinetic Model. *Water Research*, 25(9):797–805.

Amy, G.L,. P.A. Chadik, and Z. Chowdhury. 1987. Developing Models for Predicting THM Formation Potential and Kinetics. *Jour. AWWA*, 79(7):89–97.

APHA, AWWA, WPCF (American Public Health Association, American Water Works Association, Water Pollution Control Federation). 1995. *Standard Methods for the Examination of Water and Wastewater.* 19th ed. Washington, D.C.: APHA.

Arbuckle, T.E., S.E. Hrudey, S.W. Krasner, J.R. Nuckols, S.D. Richardson, P.C. Singer, P. Mendola, L. Dodds, C. Weisel, D.L. Ashley, K.L. Froese, R.A. Pegram, I.R. Schultz, J. Reif, A.M. Bachand, F.M. Benoit, M. Lynberg, C. Poole, and K. Waller. 2002. Assessing Exposure in Epidemiologic Studies to Disinfection By-Products in Drinking Water: Report from an International Workshop. *Environ. Health Perspect.*, 110(1):53–60

Arora, H., M. LeChevallier, and K.L. Dixon. 1997. DBP Occurrence Survey. *Jour. AWWA*, 89(6):60–68.

Beckwith, R.C., T.X. Wang, and D.W. Margerum. 1996. Equilibrium and Kinetics of Bromine Hydrolysis. *Inorg. Chem.*, 35(4):995–1000.

Black, B.D., G.W. Harrington, and P.C. Singer. 1996. Reducing Cancer Risks by Improving Organic Carbon Removal. *Jour. AWWA*, 88(6):40–52.

Boyce, S.D., and J.F. Hornig. 1983. Reaction Pathways of Trihalomethane Formation from the Halogenation of Dihydroxyaromatic Model Compounds for Humic Acid. *Environ. Sci. Technol.*, 17:202–211.

Bruchet, A., C. Rousseau, and J. Mallevialle. 1990. Pyrolysis GC-MS for Investigating High Molecular Weight THM Precursors and Other Refractory Organics. *Jour. AWWA*, 82(9):66–74.

Capellos, Ch., and B.H.J. Bielski. 1972. *Kinetic Systems.* New York: Wiley-Interscience.

Carlson, M., and D. Hardy. 1998. Controlling DBPs with Monochloramine. *Jour. AWWA*, 90(2):95–106.

Casey, T.J., and K.H. Chua. 1997. Aspects of THM Formation in Drinking Water. *J. Water Sup. Res. Technol.-AQUA*, 46(1):31–39.

Castro, C.E., S.K O'Shea, and W. Wang. 1996. Biodehalogenation: Oxidative and Hydrolytic Pathways in the Transformations of Acetonitrile, Chloroacetonitrile, Chloroacetic Acid, and Chloroacetamide by Methylosinus Trichosporium OB-3b. *Environ. Sci. Technol.*, 30:1180–1185.

Chaidou, C.I., V.I. Georgakilas, C. Stalikas, M. Saraci, and E.S. Lahaniatis. 1999. Formation of Chloroform by Aqueous Chlorination of Organic Compounds. *Chemosphere*, 39(4):587–594.

Chellam, S., and S.W. Krasner. 2001. Disinfection By-Product Relationships and Speciation in Chlorinated Nanofiltered Waters. *Environ. Sci. Technol.*, 35(19):3988–3999.

Chin, Y.P., G. Aiken, and E. O'Loughlin. 1994. Molecular Weight, Polydispersity and Spectroscopic Properties of Aquatic Humic Substances. *Environ. Sci. Technol.*, 28(11):1853–1858.

Christman, R.F., D.L. Norwood, Y. Seo, and F.H. Frimmel. 1989. Oxidative Degradation of Humic Substances from Freshwater Environments. In *Humic Substances II*. New York: John Wiley & Sons.

Clark, R.M., H. Pourmoghaddas, L.J. Wymer, and R.C. Dressman. 1996. Modeling the Kinetics of Chlorination By-Products Formation: The Effects of Bromide. *J. Water Sup. Res. Technol.-AQUA*, 45(3):112–119.

Clark, R.M., R.C. Thurnau, M. Sivaganesan, and P. Ringhand. 2001. Predicting the Formation of Chlorinated and Brominated By-Products. *J. Env. Engineering—ASCE*, 127(6):493–501.

Cowman, G.A. and P.C. Singer. 1996. Effect of Bromide Ion on Haloacetic Acid Speciation Resulting from Chlorination and Chloramination of Aquatic Humic Substances. *Environ. Sci. Technol.*, 30(1):16–24.

Croué, J. P., G.V. Korshin, and M.M. Benjamin. 2000. *Characterization of Natural Organic Matter in Drinking Water.* Denver Colo.: AwwaRF.

Croué, J.P., and D.A. Reckhow. 1989. Destruction of Chlorination By-Products by Sulfite. *Environ. Sci. Technol.*, 23:1412–1419.

Diehl, A.C., G.E. Speitel, J.M. Symons, S.W. Krasner, C.J. Hwang, and S.E. Barrett. 2000. DBP Formation During Chloramination. *Jour. AWWA*, 92(6):76–90.

Dotson, D.A., G.R. Helz, and G.R. Sugam. 1986. Mineralization of Organic Matter and Other Chemical Effects of Chlorination. *Water Research*, 20(8):1031–1039.

Dowty, B., D. Carlisle, J.L. Laseter, and J. Storer. 1975. Halogenated Hydrocarbons in New Orleans Drinking Water and Blood Plasma. *Science*, 187(4171):75–77.

Dycus, P.J.M., K.D. Healy, G.K. Stearman, and M.J.M. Wells. 1995. Diffusion Coefficients and Molecular Weight Distributions of Humic and Fulvic Acids Determined by Flow Field Fractionation. *Separation Sci. Technol.*, 30:1435–1453.

Edzwald, J.K. 1993. Coagulation in Drinking Water Treatment. Particles, Organics and Coagulants. *Water Sci. Technol.*, 27(11):21–35.

Edzwald, J.K., W.C. Becker, and K.L. Wattier. 1985. Surrogate Parameters for Monitoring Organic Matter and THM Precursors. *Jour. AWWA*, 77(4):122–132.

El Dib, M.A., and R.K. Ali. 1995. THMs Formation During Chlorination of Raw Nile River Water. *Water Research*, 29(1):375–378.

El Din, A.M.S., R.A. Arain, and A. Ahammoud. 1998. Kinetics of Hydrolysis of Chloroform and Bromoform in Aqueous Solutions. *Desalination*, 120(1/2):41–51.

El-Shafy, A., and A. Grunwald. 2000. THM Formation in Water Supply in South Bohemia, Czech Republic. *Water Research,* 34(13):3453–3459.

Elshorbagy, W.A. 2000. Kinetics of THM Species in Finished Drinking Water. *J. Water Res. Planning Manage.-ASCE*, 126(1):21–28.

Engelholm, B.A., and G.L. Amy. 1983. A Predictive Model for Chloroform Formation from Humic Acid. *Jour. AWWA*, 75(8):418–423.

Fogelman, K.D., D.M. Walker, and D.W. Margerum. 1989. Non-metal Redox Kinetics—Hypochlorite and Hypochlorous Acid Reactions With Sulfite. *Inorg. Chem.*, 28(6):986–993.

Frimmel, F.H., S. Hesse, and G. Kleiser. 2000. Technology-Related Characterization of Hydrophilic Disinfection By-Products in Aqueous Samples. In *Natural Organic Matter and Disinfection By-Products. Characterization and Control in Drinking Water.* Edited by S.E. Barrett, S.W. Krasner, and G.L. Amy. Washington. D.C.: American Chemical Society.

Gallard, H., and U. von Gunten. 2002. Chlorination of Natural Organic Matter: Kinetics of Chlorination and of THM Formation. *Water Research*, 36(1):65–74.

Gang, D.C., T.E. Clevenger, and S.K. Banerji. 2003. Relationship of Chlorine Decay and THMs Formation to NOM Size. *J. Hazard. Mater.*, 96(1):1–12.

Gang, D.D., R.L. Segar, T.E. Clevenger, and S.K. Banerji. 2002. Using Chlorine Demand to Predict TTHM and HAA9 Formation. *Jour. AWWA*, 94(10):76–86.

Golfinopoulos, S.K., and G.B. Arhonditsis. 2002. Multiple Regression Models: A Methodology for Evaluating Trihalomethane Concentrations in Drinking Water From Raw Water Characteristics. *Chemosphere*, 47(9):1007–1018.

Gonzalez, A.C., T.M. Olson, and L.M. Rebenne. 1996. Aqueous Chlorination Kinetics and Mechanisms of Substituted Dihydrobenzenes. In *Water Disinfection and Natural Organic Matter.* Chapter 4. Edited by R.A. Minear and G. Amy. Washington, D.C.: American Chemical Society. ACS Symposium Series, 649:48–62.

Harrington, G.W., A. Bruchet, D. Rybacki, and P.C. Singer. 1996. Characterization of Natural Organic Matter and Its Reactivity with Chlorine. In *Water Disinfection and Natural Organic Matter: Characterization and Control.* Edited by R.A. Minear, and G.L. Amy. Washington, D.C.: American Chemical Society. ACS Symposium Series, 649:138–158.

Harrington, G.W., Z.K. Chowdhury, and D.M. Owen. 1992. Developing a Computer Model to Simulate DBP Formation During Water Treatment. *Jour. AWWA*, 84(11):78–87.

Heller-Grossman, L., J. Manka, and B. Limoni-Relis. 1993. Formation and Distribution of Haloacetic Acids, THM and TOX in Chlorination of Bromide-Rich Lake Water. *Water Research,* 27:1323–1330.

Hi-Tech Ltd. 2003. Hi-Tech Scientific Worldwide Markets—Products [Online]. Available: <http://www.hi-techsci.com/qf.htm>. [cited December 1, 2003]

Hi-Tech Ltd. 2003. Hi-Tech Scientific Worldwide Markets—Techniques [Online]. Available: <http://www.hi-techsci.com/techniques/qf-tech.htm>. [cited December 1, 2003]

Huang, W.J., and H.H. Yeh. 1997. The Effect of Organic Characteristics Land Bromide on Disinfection By-Products Formation by Chlorination. *J. Environ. Sci. Health Part A-Environ. Sci. Eng. Toxic Hazard. Subst. Control.*, 32(8):2311–2336.

Ivancev-Tumbas, I., B. Dalmacija, Z. Tamas, and E. Karlovic. 1999. The Effect of Different Drinking Water Treatment Processes on the Rate of Chloroform Formation in the Reactions of Natural Organic Matter with Hypochlorite. *Water Research*, 33(18):3715–3722.

Jadas-Hecart, A., A. El Morer, M. Stitou, P. Bouillot, and B. Legube. 1992. Modelization de la Demande en Chlore d'une Eau Traitée. *Water Research*, 26(8):1073–1084.

Jimenez, M.C.S., A.P. Dominguez, and J.M.C. Silverio. 1993. Reaction Kinetics of Humic Acid With Sodium Hypochlorite. *Water Research*, 27(5):815–820.

Karanfil, T., M.A. Schlautman, and I. Erdogan. 2002. Survey of DOC and UV Measurement Practices With Implications for SUVA Determination. *Jour. AWWA*, 94(12):68–80.

Kitis, M., T. Karanfil, J.E. Kilduff, and A. Wigton. 2001. The Reactivity of Natural Organic Matter to Disinfection By-Products Formation and Its Relation to Specific Ultraviolet Absorbance. *Water Sci. Technol.*, 43(2):9–16.

Kitis, M, T. Karanfil, A. Wigton, and J.E. Kilduff. 2002. Probing Reactivity of Dissolved Organic Matter for Disinfection By-Product Formation Using XAD-8 Resin Adsorption and Ultrafiltration Fractionation. *Water Research*, 36(15):3834–3848.

Korshin, G.V., M.M. Benjamin, O. Hemingway, and W. Wu. 2002b. *Development of Differential UV Spectroscopy for On-line DBP Monitoring.* Denver, Colo.: AwwaRF and AWWA.

Korshin, G.V., M.M. Benjamin, and R.S. Sletten. 1997. Adsorption of Natural Organic Matter on Iron Oxide. Effects of NOM Composition and on the Formation of Organo-Halide Compounds. *Water Research*, 31:1643–1650.

Korshin, G.V., M.M. Benjamin, and H-B. Xiao. 2001. Interactions of Chlorine With Natural Organic Matter and Formation of Intermediates: Evidence by Differential Spectroscopy. *Acta Hydroch. Hydrob.*, 28(7):378–384.

Korshin, G.V., and M.D. Jensen. 2001. Electrochemical Reduction of Haloacetic Acids and Exploration of Their Removal by Electrochemical Treatment. *Electrochim. Acta*, 47(5):747–751.

Korshin, G.V., C-W. Li, and M.M. Benjamin. 1996. Use of UV Spectroscopy to Study Chlorination of Natural Organic Matter. In *Water Disinfection and Natural Organic Matter.* Chapter 12. Edited by R.A. Minear and G. Amy. Washington, D.C.: American Chemical Society. ACS Symposium Series, 649:182–195.

Korshin, G.V., C-W. Li, and M.M. Benjamin. 1997a. The Decrease of UV Absorbance as an Indicator of TOX Formation. *Water Research*, 31(4):946–949.

Korshin, G.V., C-W. Li, and M.M. Benjamin 1997b. Monitoring the Properties of Natural Organic Matter Through UV Spectroscopy. Evaluation of a Consistent Theory. *Water Research*, 31(7):1787–1795.

Korshin, G.V., C-W. Li, and M.M. Benjamin. 1999. Use of Differential Spectroscopy to Evaluate the Structure and Reactivity of Humics. *Water Sci. Technol.*, 40(9):9–16.

Korshin, G.V., W. Wu, M.M. Benjamin, and O. Hemingway. 2002a. Correlations Between Differential Absorbance and the Formation of Individual DBP Species. *Water Research*, 36(13):3273–3282.

Krasner, S.W., M.J. McQuire, J.C. Jacangelo, N.L. Patania, K.M. Reagan, and E.M. Aieta. 1989. The Occurrence of Disinfection By-Products in US Drinking Water. *Jour. AWWA*, 81(8):41–53.

Kumar, K, and D.W. Margerum. 1987. Kinetics and Mechanism of General Acid-Assisted Oxidation of Bromide by Hypochlorite and Hypochlorous Acid. *Inorg. Chem.*, 26:2706–2711.

Larson, R.A., and E.J. Weber. 1994. *Reaction Mechanisms in Environmental Organic Chemistry.* Boca Raton, Fla.: Lewis Publishers.

Leenheer, J.A. 1981. Comprehensive Approach to Preparative Isolation and Fractionation of Dissolved Organic Carbon from Natural Waters and Wastewaters. *Environ. Sci. Technol.*, 15(5):578–587.

Leenheer, J.A., M.A. Wilson, and R.L. Malcolm. 1987. Presence and Potential Significance of Aromatic Ketone Groups in Aquatic Humic Substances. *Org. Geochem.*, 11:273–280.

Leung, T.V., C.J. Peters, and R. Perry. 1982. Influence of Bromide and Ammonia Upon the Formation of THN Under Water-Treatment Conditions. *Environ. Sci. Technol.*, 16:473–479.

Li, C.W., M.M. Benjamin, and G.V. Korshin. 2000. Use of UV Spectroscopy to Characterize Reactions Between NOM and Free Chlorine. *Environ. Sci. Technol.*, 34(12):2570–2575.

Li, C.W., M.M. Benjamin, and G.V. Korshin. 2002. The Relationship Between TOX Formation and Spectral Changes Accompanying Chlorination of Pre-Concentrated or Fractionated NOM. *Water Research*, 36(13):3265–3272.

Li, C.W., G.V. Korshin, and M.M. Benjamin. 1998. Monitoring Formation of Disinfection By-Products with Differential UV Spectroscopy. *Jour. AWWA*, 90(8):88–100.

Malcolm, R.L. 1990. The Uniqueness of Humic Substances in Each Soil, Stream and Marine Environments. *Analytica Chimica Acta*, 232(1):19–30.

Martin-Mousset, B., J.P. Croué, E. Lefebvre, and B. Legube. 1997. Distribution and Characterization of Dissolved Organic Matter of Surface Waters. *Water Research*, 31:541–553.

Marx, J.L. 1974. Drinking Water: Another Source of Carcinogens? *Science*, 186(4166):809–811.

McClellan, J.N., D.A. Reckhow, J.E. Tobiason, J.K. Edzwald, and D.B. Smith. 2000. A Comprehensive Kinetic Model for Chlorine Decay and Chlorination By-Products Formation. In *Natural Organic Matter and Disinfection By-Products. Characterization and Control in Drinking Water.* Chapter 15. Edited by S.E. Barrett, S.W. Krasner, and G.L. Amy. Washington, D.C.: American Chemical Society. ACS Symposium Series, 761:223–246.

Miller, J.W., and P.C. Uden. 1983. Characterization of Aqueous Chlorination Products of Humic Substances. *Environ. Sci. Technol.*, 17:150–157.

Milot, J., M.J. Rodriguez, and J.B. Serodes. 2000. Modeling the Susceptibility of Drinking Water Utilities to Form High Concentrations of Trihalomethanes. *J. Environ. Manage.*, 60(2):155–171.

Milot, J., M.J. Rodriguez, and J.B. Serodes. 2002. Contribution of Neural Networks for Modeling Trihalomethanes Occurrence in Drinking Water. *J. Water Res. Planning Manage.-ASCE*, 128(5):370–376.

Montgomery Watson. 1993. *Disinfection/Disinfection By-Products Database and Model Project.* Final Report. Denver, Colo.: AWWA.

Morrow, C.M., and R.A. Minear. 1987. Use of Regression Models to Link Raw Water Characteristics to THM Concentrations in Drinking Water. *Water Research,* 21:41–49.

Nokes, C.J., E. Fenton, and C.J. Randall. 1999. Modeling the Formation of Brominated Trihalomethanes in Chlorinated Drinking Waters. *Water Research*, 33(17):3557–3568.

Novak, J.M., G.L. Mills, and P.M. Bertsch. 1992. Estimating the Percent Aromatic Carbon in Soil and Aquatic Humic Substances Using Ultraviolet Absorbance Spectroscopy. *J. Environ. Qual.,* 21:144–147.

Oliver, B.G. 1983. Dihaloacetonitriles in Drinking Water: Algae and Fulvic Acid as Precursors. *Environ. Sci. Technol.*, 17(1):80–83.

Pehkonen, S.O., R.L. Siefert, and M.R. Hoffmann. 1995. Photoreduction of Iron Oxyhydroxides and the Photooxidation of Halogenated Acetic Acids. *Environ. Sci. Technol.,* 29:1215–1222.

Peters, C.J., R.J. Young, and R. Perry. 1980. Factors Influencing the Formation of Haloforms in the Chlorination of Humic Materials. *Environ. Sci. Technol.*, 14:1391–1395.

Petri, M., M. Wilker, H.H. Stabel, and E. Gilbert. 1997. Formation of Brominated Disinfection By-Products After Chlorination of Water From Lake Constance Depending on Treatment Steps and Bromide Concentration. *Acta Hydroc. Hydrob.*, 25(6):319–328

Pomes, M.L., W.R. Green, E.M. Thurman, W.H. Orem, and H.E. Lerch. 1999. DBP Formation Potential of Aquatic Humic Substances. *Jour. AWWA*, 91(3):103–115

Powell, J.C., J.R. West, N.B. Hallam, C.F. Forster, and J. Simms. 2000. Performance of Various Kinetic Models for Chlorine Decay. *J. Water Res. Planning Manage.*, 126(1):13–20.

Qualls, R.G., and J.D. Johnson. 1983. Kinetics of the Short-Term Consumption of Chlorine by Fulvic Acid. *Environ. Sci. Technol.*, 17:692–698.

Raitt, D. 2000. The Greatest Engineering Achievements of the Twentieth Century. *Electronic Library,* 18(2):94–96.

Rathbun, R.E. 1996a. Bromine Incorporation Factors for Trihalomethane Formation for the Mississippi, Missouri, and Ohio Rivers. *Sci. Total Environ.*, 192:111–118.

Rathbun, R.E. 1996b. Regression Equations for Disinfection By-Products for the Mississippi, Ohio and Missouri Rivers. *Sci. Total Environ.*, 191:235–244.

Rebenne, L.M., A.C. Gonzalez, and T.M. Olson. 1996. Aqueous Chlorination Kinetics and Mechanism of Substituted Dihydroxybenzenes, *Environ. Sci. Technol.*, 30:2235–2242.

Reckhow, D.A., P.C. Singer, and R.L. Malcolm. 1990. Chlorination of Humic Materials: By-Product Formation and Chemical Interpretation. *Environ. Sci. Technol.*, 24:1655–1664.

Richardson, S.D., J.E. Simmons, and G. Rice. 2002. Disinfection By-Products: The Next Generation. *Environ. Sci. Technol.*, 36(9):198A–205A.

Richardson, S.D., A.D. Thruston, T.V. Caughran, P.H. Chen, T.W. Collette, and T.L. Floyd. 1999. Identification of New Drinking Water Disinfection By-Products Formed in the Presence of Bromide. *Environ. Sci. Technol.*, 33(19):3378–3383.

Rios, R.V.R.A, L.L. da Rocha, T.G. Vieira, R.M. Lago, and R. Augusti. 2000. On-Line Monitoring by Membrane Introduction Mass Spectrometry of Chlorination of Organics in Water. Mechanistic and Kinetic Aspects of Chloroform Formation. *J. Mass Spectrom.*, 35(5):618–624.

Roberts, M.G., P.C. Singer, and A. Obolensky. 2002. Comparing Total HAA and Total THM Concentrations using ICR Data. *Jour. AWWA*, 94(1):103–116

Rodriguez, M.J., J. Serodes, and M. Morin. 2000. Estimation of Water Utility Compliance with Trihalomethane Regulations Using a Modeling Approach. *J. Water Sup. Res. Technol.-AQUA*, 49(2):57–73.

Rook, J.J. 1974. Formation of Haloforms During Chlorination of Natural Waters. *J. Soc. Water Treat. Exam.*, 23:234–243.

Rook, J.J. 1977. Chlorination Reactions of Fulvic Acids in Natural Waters. *Environ. Sci. Technol.*, 11:478–482.

Simmons, J.E., S.D. Richardson, T.F. Speth, R.J. Miltner, G. Rice, K.M. Schenck, E.S. Hunter, and L.K. Teuschler. 2002. Development of a Research Strategy for Integrated Technology-Based Toxicological and Chemical Evaluation of Complex Mixtures of Drinking Water Disinfection By-Products. *Environ. Health Perspect.*, 110(6):1013–1024.

Singer, P.C., and S.D. Chang. 1989. Correlations Between Trihalomethanes and Total Organic Halides During Water Treatment. *Jour. AWWA*, 81(8):61–65.

Singer, P.C., A. Obolensky, and A. Greiner. 1995. DBPs in Chlorinated North Carolina Drinking Waters. *Jour. AWWA*, 87(10):83–92.

Solarik G., R.S. Summers, J. Sohn, W. Swanson, Z. Chowdhury, and G. Amy. 2000. Extensions and Verifications of the Water Treatment Plant Model for Disinfection By-Product Formation. In *Natural Organic Matter and Disinfection By-Products*. Edited by S. Barrett, S. Krasner, and G. Amy. ACS Symposium Series 761. Washington D.C.: American Chemical Society.

Stevens, A.A., L.A. Moore, and R.J. Miltner. 1989. Formation and Control of Non-THM Disinfection By-Products. *Jour. AWWA*, 81(10):54–60.

Stevens, A.A., C.J. Slocum, D.R. Seeger, and G.G. Robeck. 1976. Chlorination of Organics in Drinking Water. *Jour. AWWA*, 68(11):615–620.

Suffet, M., L. Brenner, and B. Silver. 1976. *Environ. Sci. Technol.*, 13:1273–1275

Sung, W., B. Reilley-Matthews, D.K. O'Day, and K. Horrigan. 2000. Modeling DBP Formation. *Jour. AWWA*, 92(5):53–63.

Symons J.M., R. Xia, G.E. Speitel Jr., A.C. Diehl, C.J. Hwang, S.W. Krasner, and S.E. Barrett. 1998. *Factors Affecting Disinfection By-Product Formation During Chloramination.* Denver, Colo.: AwwaRF and AWWA.

Tapudurti, K.V., and C.N. Haas. 1991. THM Formation by the Transfer of Active Chlorine from Monochloramine to Phloroacetophenone. *Jour. AWWA*, 83(5):62–66.

Traina, S.J., J. Novak, and N.E. Smeck. 1990. An Ultraviolet Absorbance Method of Estimating the Percent Aromatic Carbon Content in Humic Acids. *J. Environ. Qual.*, 19:151–153.

Tretyakova, N.Y., A.T. Lebedev, and V.S. Petrrosyan. 1994. Degradative Pathways for Aqueous Chlorination of Orcinol. *Environ. Sci. Technol.*, 28:606–613

Uden, P.C., and J.W. Miller. 1983. Chlorinated Acids and Chloral in Drinking Water. *Jour. AWWA*, 75(10):524–527.

Urano, K., H. Wada, and T. Takemasa. 1983. Empirical Rate Equation for THM Formation With Chlorination of Humic Substances in Water. *Water Research*, 17:1797–1802.

USEPA (U.S. Environmental Protection Agency). 1995a. Method 551.1. Determination of Chlorination Disinfection Byproducts, Chlorinated Solvents, and Halogenated Pesticides/Herbicides in Drinking Water by Liquid–Liquid Extraction and Gas Chromatography with Electron-Capture Detection. In *Methods for the Determination of Organic Compounds in Drinking Water: Supplement III*. Cincinnati, Ohio: USEPA.

USEPA (U.S. Environmental Protection Agency). 1995b. Method 552.2. Determination of Haloacetic Acids and Dalapon in Drinking Water by Liquid–Liquid Extraction, Derivatization and Gas Chromatography With Electron-Capture Detection. In *Methods for the Determination of Organic Compounds in Drinking Water: Supplement III*. Cincinnati, Ohio: USEPA.

Wang, T.X., M.D. Kelley, J.N. Cooper, R.C. Beckwith, and D.W. Margerum. 1994. Equilibrium, Kinetic and UV Spectral Characteristics of Aqueous Bromine Chloride, Bromine and Chlorine Species. *Inorg. Chem.*, 33(25):5872–5878.

Wang, T.X., and D.W. Margerum. 1994. Kinetics of Reversible Chlorine Hydrolysis. Temperature Dependence and General Acid Base Assisted Mechanisms. *Inorg. Chem.,* 33(6):1050–1055.

Wang, Z.D., B.C. Pant, and C.H. Langford. 1990. Spectroscopic and Structural Characterization of a Laurentian Fulvic Acid: Notes on the Origin of Color. *Anal. Chim. Acta*, 232:43–49.

Westerhoff, P., J. Debroux, G.L. Amy, D. Gatel, V. Mary, and J. Cavard. 2000. Applying DPB Models to Full-Scale Plants. *Jour. AWWA*, 92(3):89–102.

Westerhoff, P, R.G. Song, G. Amy, and R. Minear. 1998. Numerical Kinetic Models for Bromide Oxidation to Bromine and Bromate. *Water Research*, 32:1687–1699.

Wu, W.W., and P.A. Chadik. 1998. Effect of Bromide Ion on Haloacetic Acid Formation During Chlorination of Biscayne Aquifer Water. *J. Environ. Eng.-ASCE*, 124(10):932–938.

Zhang, X.R., and R.A. Minear. 2002. Decomposition of Trihaloacetic Acids and Formation of the Corresponding Trihalomethanes in Drinking Water. *Water Research*, 36(14):3665–3673.

Traina, S.J., J. Novak and N.E. Smeck. 1990. An Ultraviolet Absorbance Method of Estimating the Percent Aromatic Carbon Content in Humic Acids. *J. Environ. Qual.*, 19:151–153.
Tretiakova, N.Y., A.T. Lebedev and V.S. Petrosyan. 1994. Degradative Pathways for Aqueous Chlorination of Orcinol. *Environ. Sci. Technol.*, [illegible]
Uden, P.C. and J.W. Miller. 1983. Chlorinated Acids and Chloral in Drinking Water. *Jour. AWWA*, 75(10):524–527.
Urano, K., H. Wada and T. Takemasa. 1983. Empirical Rate Equation for THM Formation With Chlorination of Humic Substances in Water. *Water Research*, 17(12):1797–1802.
USEPA (U.S. Environmental Protection Agency). 1995a. Method 551.1. Determination of Chlorination Disinfection Byproducts, Chlorinated Solvents, and Halogenated Pesticides/Herbicides in Drinking Water by Liquid-Liquid Extraction and Gas Chromatography with Electron-Capture Detection. In *Methods for the Determination of Organic Compounds in Drinking Water: Supplement III*. Cincinnati, Ohio: USEPA.
USEPA (U.S. Environmental Protection Agency). 1995b. Method 552.2. Determination of Haloacetic Acids and Dalapon in Drinking Water by Liquid-Liquid Extraction, Derivatization and Gas Chromatography With Electron Capture Detection. In *Methods for the Determination of Organic Compounds in Drinking Water: Supplement III*. Cincinnati, Ohio: USEPA.
Wang, T.X., M.D. Kelley, J.N. Cooper, R.C. Beckwith, and D.W. Margerum. 1994. Equilibrium, Kinetic and UV Spectral Characteristics of Aqueous Bromine Chloride, Bromine and Chlorine Species. *Inorg. Chem.*, 33(25):5872–5878.
Wang, T.X. and D.W. Margerum. 1994. Kinetics of Reversible Chlorine Hydrolysis: Temperature Dependence and General-Acid/Base-Assisted Mechanisms. *Inorg. Chem.*, [illegible]
Wershaw, R.L., [illegible] 1990. Spectroscopic and Structural Characterization [illegible]
[illegible]

ABBREVIATIONS

A	absorbance
ACS	American Chemical Society
AOX	adsorbable organic halogen
APHA	American Public Health Association
A_{272}	absorbance at 272 nm
AWWA	American Water Works Association
AwwaRF	Awwa Research Foundation
BCAA	bromochloroacetic acid
BCAN	bromochloroacetonitrile
BDCAA	bromodichloroacetic acid
°C	degrees centigrade
Cl_2:DOC	chlorine to DOC mass ratio
CH	chloral hydrate
cm	centimeter
Co.	county
DBAA	dibromoacetic acid
DBAN	dibromoacetonitrile
DBCAA	dibromochloroacetic acid
DBP	disinfection by-product
DCAA	dichloroacetic acid
DCAN	dichloroacetonitrile
DCP	1,1-dichloro-2-propanone
ΔA	differential absorbance
ΔA_{272}	differential absorbance measured at 272 nm
ΔA_{λ}	differential absorbance measured at wavelength
DOC	dissolved organic carbon
DPD	N,N'-diethyl-1,4-phenylenediamine
ECD	electron capture detector
GC	gas chromatograph
HAA	haloacetic acid
HAA5	total of five haloacetic acids
HAN	haloacetonitrile
HK	haloketone
HPLC	high-performance liquid chromatography
i.d.	inner diameter

L	liter
LW	Lake Washington
m	meter
M	moles per liter
MBAA	monobromoacetic acid
MCAA	monochloroacetic acid
MDL	method detection limit
mg/L	milligram per liter
min	minute
mL	milliliter
mm	millimeter
MΩ	mega-ohm
MTBE	methyl-tert-butyl ether
MWD	metropolitan water district
µg/L	microgram per liter
ng	nanogram
nm	nanometer
NOM	natural organic matter
PEEK	poly-ethyl-ethyl-ketone
PHA	polyhydroxyaromatic moiety
PTFE	polytetrafluoroethylene
R2	regression coefficient
s	second
$SUVA_{254}$	specific absorbance at 254 nm
t	time
TBAA	tribromoacetic acid
TCAA	trichloroacetic acid
TCAN	trichloroacetonitrile
TCP	1,1,1-trichloro-2-propanone
THM	trihalomethane
THMFP	trihalomethane formation potential
TOX	total organic halogen
TTHMs	total of trihalomethanes
USEPA	United States Environmental Protection Agency
UV	ultraviolet
UV/VIS	ultraviolet/visible
UW	University of Washington
WPCF	Water Pollution Control Federation
WTP	water treatment plant

IWA Publishing's authorised EU representative for General Product Safety Regulations is Diane D'Arras, 15 rue Duret, 75116 Paris, France, e-mail: safety@iwap.co.uk.

Printed and bound by CPI Group (UK) Ltd, Croydon, CR0 4YY
17/09/2026
02218798-0019